T0296580

HYDROSTATICS

HYDROSTATICS

A Text-Book for the use of
First Year Students at the Universities and
for the Higher Divisions in Schools

by
A. S. RAMSEY, M.A.

CAMBRIDGE
AT THE UNIVERSITY PRESS
1961

CAMBRIDGE
UNIVERSITY PRESS

University Printing House, Cambridge CB2 8BS, United Kingdom

Cambridge University Press is part of the University of Cambridge.

It furthers the University's mission by disseminating knowledge in the pursuit of education, learning and research at the highest international levels of excellence.

www.cambridge.org
Information on this title: www.cambridge.org/9781316633359

© Cambridge University Press 1961

First edition 1936
Second edition 1946
Reprinted 1947, 1951, 1956, 1961
First paperback edition 2016

A catalogue record for this publication is available from the British Library

ISBN 978-1-316-63335-9 Paperback

PREFACE

This book is a companion volume to my books on *Dynamics* and *Statics* published in recent years. But while they were intended primarily for the Higher Divisions in Schools, this book is primarily intended for First Year Students in the Universities; because Hydrostatics does not appear as yet to be studied much in Schools. The greater part of this book will, however, be found to be within the capacity of boys and girls who are up to scholarship standard, and it might well be read by such of them as desire to broaden the basis of their knowledge of Mechanics before coming to the University.

The book might have been called '*Elementary Hydrostatics*' to distinguish it from the more advanced book on the subject which forms Part I of *A Treatise on Hydromechanics* by the late Dr Besant and myself, for it constitutes a simpler course and does not cover the same ground; but the title 'Elementary Hydrostatics' is too reminiscent of the days when Hydrostatics and Heat were compulsory subjects for all candidates for the Ordinary Degree at Cambridge.

Some acquaintance with the Calculus is assumed and the book covers those parts of the subject usually included in a First Year Honours course, namely, the pressure of heavy fluids, centres of pressure, floating bodies, simple applications of the metacentre and of the general pressure equations, pressure of gases, the atmosphere, hydrostatic machines and, because no book on the subject can be considered complete without it, a short chapter on Capillarity.

The Examples have been chosen for the most part from papers set in the Mathematical Tripos and in College and Inter-collegiate Examinations, their sources being indicated by the letters M. T., C. and I.

My thanks are tendered to Dr S. Verblunsky for much valuable criticism and help in reading the proofs and verifying Examples and to the staff of the University Press for their careful printing and reading. There are doubtless some residual errors, and I shall be grateful to any readers who will help me to correct them.

A. S. R.

November 1935

CONTENTS

Chapter I: INTRODUCTION

Chapter II: PRESSURE OF HEAVY FLUIDS

Chapter III: CENTRES OF PRESSURE

Chapter IV: THRUSTS ON CURVED SURFACES. FLOATING BODIES

Chapter V: STABILITY OF FLOATING BODIES

HYDROSTATICS

Chapter I

INTRODUCTION

1·1. Hydrostatics is the branch of mathematics which is concerned with the conditions under which systems of forces maintain masses of fluid in a state of relative equilibrium, together with the determinations of the reactions exerted by fluids upon bodies with which they are in contact.

1·11. Shearing stress. At any point P inside a material substance or 'medium' imagine a small plane surface A to be drawn lying wholly within the medium. The surface A may be regarded as dividing or separating the medium at P; and, since action and reaction are equal and opposite, the two portions of the

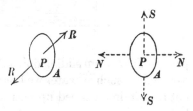

medium on opposite sides of the surface A exert on one another equal and opposite forces R. Without much more data we cannot specify the direction of the force R, but in general it can be resolved into a component N normal to A and a component S tangential to A. If for convenience we take the area of A to be the unit of area, then the force R is the **stress per unit area** across A; the normal component N is the **thrust or tension per unit area** across A according to the sense in which it acts, and the tangential component S is the **shearing stress per unit area** along A.

1·12. A fluid *is a substance which flows or is capable of flowing.* The two most familiar forms of fluid are air and water. The property of these substances which distinguishes them most readily from solids is the ease with which any masses of them can be subdivided. They offer practically no resistance to separation, or, in mathematical language, they do not exert shearing stress.

Let ABC represent a body on a horizontal plane AC, and let DE be a hypothetical oblique plane section. Suppose the body to be a solid in equilibrium. Considering the portion DBE, its weight can be resolved into components in the direction ED and at right angles to ED, and these forces are balanced by the stresses across the plane ED which, as in 1·11, consist of a shearing stress along DE and a normal thrust. Consequently it is the shearing stress in the plane DE which prevents the

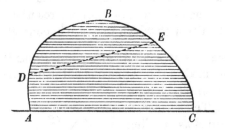

portion DBE from sliding down the plane ED. But if the body were a fluid such as water we know from experience that it will not remain in a heaped up position, and the reason for this is because in equilibrium a fluid cannot exert shearing stress or resist indefinitely any force of that nature.

This fundamental property of a fluid provides another **definition**: viz. *a fluid is a substance which will yield to any continued shearing stress however small.*

This definition embraces substances as widely different as water, treacle and coal tar. To explain the difference between such substances mathematically, it must be observed that when a fluid is *in motion* there are in general frictional forces between its particles; the magnitude of such forces depends on the *viscosity* or 'stickiness' of the fluid. The viscosity manifests itself in fluid motion in the production of shearing stresses which check the free motion of the particles. Thus reverting again to the figure, the external force on DBE tends to produce a shearing stress in the plane DE, and if ABC were a mass of non-viscous fluid such as water it would yield instantly to the externally applied tangential forces and subside on to the horizontal plane; if it were a mass of viscous fluid such as treacle the subsidence would take longer, and if it were a mass of very viscous fluid such as coal tar it would take a very long time to subside. But we include coal tar among the class of fluids because it will yield to any shearing stress *however small* provided that stress acts continuously for a sufficient time. And this property serves to distinguish *viscous fluids* from *plastic solids* such a putty, in that, with the latter class of substances, a stress of a *definite magnitude* is required to produce a deformation; while in the former class it is only necessary to have a certain duration of time for any shearing stress however small to produce a permanent displacement.

It must be remembered that viscosity or 'stickiness' is only effective in producing shearing stresses when a fluid is in motion, and consequently *there are no shearing stresses in a fluid in equilibrium*. This means that if the medium in 1·11 is a fluid in equilibrium, then at every point and however the surface A may be orientated the component S is always zero, and the thrust on any plane surface in the fluid is at right angles to it.

It follows that, in a continuous mass of fluid in equilibrium, *any* portion that we select, no matter how large or small, is in equilibrium under the action of external forces and the normal pressures of the surrounding fluid upon it. And as there are no shearing stresses the resolved parts of these forces in any direction must balance one another.

1·13. A perfect fluid is an ideal substance which, whether at rest or in motion, is incapable of exerting or offering resistance to shearing stress.

1·14. Liquids and gases. Fluids are of two kinds:

(i) **Liquids**, which are incompressible or nearly so.

(ii) **Gases**, which are easily compressible and can be made to expand indefinitely by increasing the volume to which they have access.

Water, for example, is not absolutely incompressible, but it requires such a great pressure to produce a very small relative diminution in volume that for most practical purposes it may be regarded as incompressible.

The distinction between solid, liquid and gas is sometimes stated in this way: a solid of definite mass has size and shape; a definite mass of liquid has size but no shape, and a definite mass of gas has neither size nor shape.

All gases can be converted into liquids by sufficient lowering of temperature and increase of pressure. For each gas there is a *critical temperature* such that for higher temperatures no in-crease of pressure however great will cause the gas to condense, while for lower temperatures the gas can be condensed by in-creasing the pressure. At temperatures below the critical temperature the gas is called a *vapour* and when above the critical temperature it is called a *permanent gas*.

1·2. Pressure. We have used the word 'pressure' without defining it. The idea of pressure implies something pressed upon, e.g. a surface subject to a certain force or thrust. When the thrust upon every portion of a plane surface is proportional to the area of the portion there is said to be a *uniform pressure* on the surface; and *the measure of the pressure is the thrust upon a unit of area.*

The **mean pressure** on a given plane surface is the uniform pressure which would produce the same resultant thrust as the actual pressure produces.

The **pressure at a point** of a surface is the limit of the mean pressure on a small area surrounding the point as the area is diminished indefinitely.

We may define **the pressure at a point of a fluid** explicitly thus: Imagine a small plane surface containing α units of area to be placed at the point (as in **1·11**) and let N denote the normal thrust of the fluid on either side of this surface, then the limit of (N/α) as $\alpha \to 0$ is the pressure at the point.

It will be noticed that this definition tacitly assumes that the result $\lim_{\alpha \to 0} (N/\alpha)$ is independent of the orientation of the small plane surface, and we must now proceed to justify that assumption.

1·21. The pressure at a point in a fluid in equilibrium is the same in every direction.

In the fluid draw a small tetrahedron $OABC$ with three mutually perpendicular faces OBC, OCA, OAB. We suppose that the fluid is subject to an external field of force (such as gravity), the effect of which can be measured as so much force per unit mass of fluid. Then the component parallel to OA of the external forces on the fluid in the tetrahedron may be denoted by

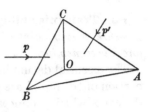

$$X \times \text{mass of fluid in } OABC,$$

or by

$$\rho X \times \text{volume } OABC,$$

where ρ is the mean density of the fluid in $OABC$.

Now let p, p' denote the mean pressures on the surfaces OBC, ABC, so that the actual thrusts on these surfaces are $p \cdot \triangle OBC$ and $p' \cdot \triangle ABC$; and let θ denote the angle between OBC and ABC.

By considering the equilibrium of the fluid in the tetrahedron and resolving parallel to OA, we get

$$p \cdot \triangle OBC - p' \cdot \triangle ABC \cos\theta + \rho X (\text{vol. } OABC) = 0 \quad (1),$$

or

$$(p - p') \triangle OBC + \tfrac{1}{3}\rho X \cdot OA \cdot \triangle OBC = 0,$$

or

$$p - p' + \tfrac{1}{3}\rho X \cdot OA = 0 \quad \dots\dots\dots\dots\dots(2).$$

Now let the tetrahedron diminish in size to vanishing point, then p, p' become the pressure at O in two different arbitrarily assigned directions, and since $OA \to 0$, the equation reduces to $p = p'$, which proves the proposition.

1·211. In the case of a fluid in motion, provided that the fluid is the ideal fluid of **1·13**, so that there are no shearing stresses, the foregoing proposition is still true, for in this case instead of the statical equation (1) we shall have a dynamical equation which will only differ from (1) by the addition of a term representing 'mass × acceleration'; i.e. a term like the term in X, but with acceleration instead of X, and this term will give rise to a term in (2) which also tends to zero as the tetrahedron tends to vanishing point.

1·22. Dimensions of pressure. Since pressure is force per unit area, its dimensions in terms of the fundamental units of mass, length and time are

$$MLT^{-2}/L^2 = ML^{-1}T^{-2}.$$

1·3. Transmissibility of liquid pressure. *An increase of pressure at any point of a liquid at rest under given external forces is transmitted to every other point of the liquid.*

The essential datum here is that the external forces on any portion of the liquid are prescribed and remain unaffected by the cause which produces the change of pressure.

Let A, B be any two points in the liquid. About the straight line AB describe a thin cylinder with plane ends at right angles

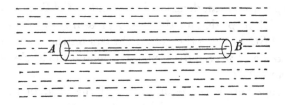

to AB. Then assuming this cylinder to lie wholly in the liquid, the difference of the thrusts on its ends must be equal to the resolved part in the direction AB of the external forces on the liquid in the cylinder. But this is constant, and therefore the difference between the pressures at A and B must be constant, so that any increase of pressure at A must be accompanied by a like increase at B.

When the straight line AB does not lie wholly in the liquid, let the points A, B be joined by a succession of lines AC, CD ... KB which do lie in the liquid; then it can be shewn as above that the pressure difference between each pair of points remains constant.

1·31. Hydraulic or Bramah's press. In outline this apparatus consists of two vertical cylinders communicating with one another by a tube near their bases, and each fitted with a piston, the space below the pistons being filled with water.

If a downward force P be applied to the piston of area A this creates an additional pressure P/A per unit area throughout the liquid, and consequently if the second piston is of area B the resulting upward thrust upon it is represented by $Q = BP/A$; so that

$$Q : P = B : A,$$

or the thrusts on the pistons are proportional to their areas.

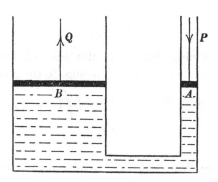

Hence by adjusting the ratio of the areas we arrive at the 'hydro-static paradox', that *any force however small may by its transmission through a liquid be made to support any weight however large.*

The apparatus may be used as a press if the load to be compressed is placed upon the piston B and pressed against a fixed barrier above the piston. In practice cupped leather collars round the pistons are needed to prevent oozing of the water. They are so placed that the water presses the collar against the sides of both the cylinder and the piston. A portion of such a collar is shewn in section in the figure.

1·4. Compression. Let a change of volume take place in a given mass of fluid, and let V, V' denote the initial and final volumes. We may define the **compression** as the ratio of the reduction in volume to the original volume; i.e.

$$\text{compression} = (V - V')/V,$$

or, in the notation of the Calculus,

$$= -dv/v.$$

The **compressibility** of a fluid is the limiting ratio of the compression to the increase of pressure which produces it; or, if P, P' denote initial and final pressures,

$$\text{compressibility} = \lim \left(\frac{V-V'}{V}\right) \Big/ (P'-P)$$

$$= -\frac{1}{v}\frac{dv}{dp}.$$

Similarly we may define the **elasticity** of a fluid as the limiting ratio of the increase of pressure to the compression it produces; i.e.

$$\text{elasticity} = \lim (P'-P) \Big/ \left(\frac{V-V'}{V}\right)$$

$$= -v\frac{dp}{dv}.$$

1·5. Density. The density of a substance is the mass of a unit of volume of the substance; e.g. the number of pounds per cubic foot or the number of grammes per cubic centimetre.

The **specific gravity** of a substance is the ratio of the density of that substance to the density of a standard substance.

The substance usually adopted as the standard is pure water. In rough calculations the mass of a cubic foot of water may be taken as 1000 oz. or 62·5 lb.; but at a temperature of 4° C., i.e. the temperature at which water has its maximum density, the density is approximately 62·425.

In the metric system the density of water is 1, the gramme having been chosen as the mass of 1 cubic centimetre of pure water at 4° C.

If W denotes the weight in absolute units (poundals) of a body of mass M (lb.), then, as in books on Dynamics,

$$W = Mg,$$

where g is the acceleration due to gravity.

But if the body is of density ρ and contains V units of volume, then from the definition of density

$$M = \rho V,$$

so that $\qquad\qquad W = g\rho V$ ....(1).

This relation expresses the weight in poundals, when V is measured in cubic feet and ρ is the mass in pounds of 1 cu. ft.

1·51. It will be observed that the dimensions of density in terms of the fundamental units are expressed by the formula ML^{-3}, since it denotes mass per unit volume. On the other hand specific gravity is of no dimensions, but is merely a number representing how many times as heavy as a standard substance a given substance is.

Consequently if w denotes the weight of a unit of volume of the standard substance, then the weight of a unit of volume of a substance of specific gravity S is Sw; and the weight W of V units of volume of the substance of specific gravity S is given by the formula

$$W = VSw \qquad \ldots\ldots\ldots\ldots\ldots\ldots(1).$$

1·52. Specific gravity of mixtures. Let volumes V_1, V_2, V_3 ... of fluids of specific gravities S_1, S_2, S_3 ... be mixed together and let $\bar{S}$ be the specific gravity of the mixture. Then using **1·51** (1) and assuming the volume of the mixture to be the sum of the volumes of its constituents, the weight of the mixture is

$$(V_1 + V_2 + V_3 + \ldots)\bar{S}w;$$

but this must be equal to the sum of the weights of the constituents, viz.

$$V_1 S_1 w + V_2 S_2 w + V_3 S_3 w + \ldots.$$

Whence, equating and dropping the factor w, we get

$$\bar{S} = \frac{V_1 S_1 + V_2 S_2 + V_3 S_3 + \ldots}{V_1 + V_2 + V_3 + \ldots} \qquad \ldots\ldots\ldots\ldots(1).$$

If for any reason the mixture has a volume U, not equal to the sum of the volumes of the constituents, the corresponding formula is clearly

$$\bar{S} = (V_1 S_1 + V_2 S_2 + V_3 S_3 + \ldots)/U \qquad \ldots\ldots\ldots(2).$$

Secondly, if given weights W_1, W_2, W_3 ... of the fluids are mixed, then from **1·51** (1) the volume of the mixture is $(W_1 + W_2 + W_3 + \ldots)/\bar{S}w$, and, assuming there to be no change in total volume, this must be equal to the sum of the volumes of the constituents, viz.

$$\frac{W_1}{S_1 w} + \frac{W_2}{S_2 w} + \frac{W_3}{S_3 w} + \ldots.$$

Whence we get $\quad \bar{S} = \dfrac{W_1 + W_2 + W_3 + \cdots}{\dfrac{W_1}{S_1} + \dfrac{W_2}{S_2} + \dfrac{W_3}{S_3} + \cdots}$ ....(3),

as could have been deduced from (1) by applying **1·51** (1).

The formula can also be modified if it is known that a definite change in volume results from the mixing.

1·6. Examples. (i) *Shew that the specific gravity of a mixture of n fluids is greater when equal volumes are taken than when equal weights are taken, assuming no change in volume as the result of mixing.*

Let $\bar{S}, \bar{S}'$ denote the specific gravity of the mixture according as we mix equal volumes or equal weights. Then if $s_1, s_2 \ldots s_n$ denote the specific gravities of the constituents, from **1·52** (1) and (3) we have

$$\bar{S} = \frac{1}{n}(s_1 + s_2 + \cdots + s_n) \quad \text{and} \quad \frac{1}{\bar{S}'} = \frac{1}{n}\left(\frac{1}{s_1} + \frac{1}{s_2} + \cdots + \frac{1}{s_n}\right).$$

If we now apply the theorem that the arithmetic mean of n unequal positive numbers is greater than their geometric mean, we have

$$\bar{S} > \sqrt[n]{(s_1 s_2 \ldots s_n)} \quad \text{and} \quad \frac{1}{\bar{S}'} > \frac{1}{\sqrt[n]{(s_1 s_2 \ldots s_n)}}.$$

Whence it follows that $\bar{S} > \bar{S}'$.

(ii) *Pure water is added, drop by drop, to a vessel of volume V filled with a salt solution of specific gravity s, which is allowed to overflow. Find the specific gravity of the solution when a volume v of water has been poured in.*

Let σ denote the specific gravity when a volume v of water has been added, and $\sigma + d\sigma$ the specific gravity when a volume $v + dv$ has been added; i.e. dv may denote the volume of a drop of water.

The addition of a drop at this stage means that a volume dv of specific gravity 1 is added to a volume V of specific gravity σ and forms a total volume $V + dv$ of specific gravity $\sigma + d\sigma$ whereof a drop overflows. Hence, by equating the weight of the mixture to the sum of the weights of the constituents, we get

$$(V + dv)(\sigma + d\sigma) = dv + V\sigma.$$

Neglecting the product of the small quantities dv, $d\sigma$, this equation gives

$$V\,d\sigma + dv\,(\sigma - 1) = 0$$

or $\qquad\qquad\qquad \dfrac{d\sigma}{\sigma - 1} + \dfrac{dv}{V} = 0.$

Hence by integration $\qquad \log(\sigma - 1) + \dfrac{v}{V} = C.$

But when $v = 0$ then $\sigma = s$, so that $C = \log(s - 1)$, **and**

$$\log(\sigma - 1) = \log(s - 1) - \frac{v}{V}.$$

Therefore $\qquad\qquad\qquad \sigma = 1 + (s - 1)\,e^{-v/V}.$

EXAMPLES

1. The weight of an empty vessel is 3 oz.; when filled with water it is 9 oz.; and when filled with oil, 8·4 oz. Find the specific gravity of the oil.

2. The specific gravities of gold and copper are 19·3 and 8·62. Find the specific gravity of an alloy of 11 parts by volume of gold to 1 part of copper.

3. A mixture of oxygen and nitrogen is found to consist of 21 parts of oxygen to 79 of nitrogen by volume, or by weight 23 parts to 77. Compare the densities of the gases.

4. What quantity of water must be mixed with a gallon of milk to reduce its specific gravity from 1·03 to 1·02?

5. When equal volumes of two substances are mixed the specific gravity of the mixture is 6, and when equal weights of the same substances are mixed the specific gravity of the mixture is 4. Find the specific gravities of the substances.

6. What weight of water must be added to 30 oz. of a solution whose specific gravity is 1·06 so that the specific gravity of the mixture may be 1·03?

7. A cylindrical jar contains in the bottom water to a depth of 3 inches and above the water a liquid, of smaller specific gravity, to the depth of 4 inches; the water weighs 0·5 lb. and the other liquid weighs 0·4 lb. The jar is now shaken till the liquids are thoroughly mixed (without any alteration of volume). Find the specific gravity of the mixture, and shew that the work done in shaking the jar is at least $\frac{1}{30}$ ft. lb.

8. Five litres of a liquid whose specific gravity is 1·3 are mixed with 7 litres of a liquid whose specific gravity is 0·78. If the bulk of the liquid shrinks 1 per cent. on mixing, find the specific gravity of the mixture.

9. The pressure of steam in a boiler is 150 lb. wt. per square inch; express this in dynes per sq. cm.

[Take 1 inch = 2·5 cm., 1 lb. = 450 gm., $g = 980$ c.g.s. units.] [C.]

10. Prove that, if the elasticity of a fluid is equal to the pressure, the pressure varies inversely as the volume. [M. T.]

11. Find the relation between the pressure and volume of a fluid, if the elasticity is equal to γ times the pressure.

12. A hydraulic lift (on the principle of the Bramah's press) consists of a ram of cross-section A and a piston of cross-section B. The ram supports a weight W, and the piston is driven into the cistern with a slow uniform motion by an engine of power H. The friction at the collar of the ram is F, and at the collar of the piston is F'. Prove that the speed at which W rises is

$$BH/(BW + BF + AF').$$ [M. T.]

13. A piston is slowly pushed along a cylinder against water under pressure and to prevent leakage the piston is fitted with a leather collar of breadth l which the water pushes against the walls of the cylinder. If μ is the coefficient of friction between the leather and the cylinder and r is the radius of a section of the cylinder, prove that the fraction of the work done which is spent in overcoming friction is

$$\frac{2\mu l}{r+2\mu l}.$$ [M. T.]

14. Prove that in consequence of the friction of the collars, the efficiency of the hydraulic press is reduced to $(1-4\mu k)/(1+4\mu k')$, where k and k' denote the ratios of the heights of the larger and smaller collars to their diameters.

ANSWERS

1. 0·9. **2.** 18·41. **3.** 1817:1617. **4.** 0·5 gall.

5. $6\pm2\sqrt{3}$. **6.** $28\frac{18}{53}$ oz. **7.** $\frac{27}{35}$. **8.** $1\frac{2}{207}$.

9. 10,584,000. **11.** $pv^\gamma=$ const.

Chapter II

PRESSURE OF HEAVY FLUIDS

2·1. In a later chapter we shall obtain a general expression for the pressure in a fluid which is maintained in equilibrium by a variable field of force. In this chapter we shall consider the special case in which the field of force is uniform; for example a fluid in equilibrium under the action of gravity. This case, in which there are no external forces to be considered but the weight of the fluid, is capable of simple independent treatment.

2·2. Pressure the same at all points in a horizontal plane. *In a fluid at rest under gravity the pressure is the same at all points in the same horizontal plane.*

Let A, B be any two points in the same horizontal plane and let the straight line AB lie wholly in the fluid. Let a thin

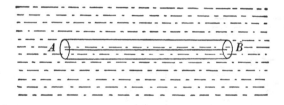

cylinder be described about the line AB as axis having plane ends at right angles to AB. Considering the equilibrium of the fluid in this cylinder, the only forces on it which have any components in the direction AB are the thrusts of the surrounding fluids on the ends at A and B.

Therefore the thrusts on the ends at A and B are equal. But the areas of the ends are equal, therefore the mean pressures on the ends are equal. Hence by diminishing the area of the cross section of the cylinder we get

$$\text{pressure at } A = \text{pressure at } B.$$

It is to be observed that this proposition involves no assumption about the density of the fluid. It is true whether the fluid be homogeneous or not, and for gases as well as for liquids. The only limitation is that it shall be possible to join the points A, B by a horizontal line lying in the fluid.

2·3. *In homogeneous fluid at rest under gravity the difference between the pressures at two points is proportional to the difference of their depths.*

Let A, B be two points in the same vertical line in the fluid. About AB as axis describe a cylinder of cross section α with plane horizontal ends at A and B. Let p, p' denote the pressures at A, B and ρ the density of the fluid.

Considering the equilibrium of the fluid in the cylinder, the vertical forces acting upon it are its weight $g\rho\alpha AB$ and the thrusts of the surrounding fluid on its ends, viz. $p\alpha$ upwards at A and $p'\alpha$ downwards at B.

Therefore $\qquad p\alpha - p'\alpha = g\rho\alpha AB$

or $\qquad\qquad p - p' = g\rho AB$ ....(1).

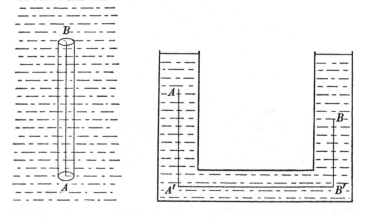

It is easy to see that the result holds good when the points A, B are not in the same vertical. If for example the fluid is contained in a vessel as depicted in the figure in which

A', B' are points in a horizontal line vertically below A, B, then

$$\text{press. at } A' - \text{press. at } A = g\rho AA'$$

and

$$\text{press. at } B' - \text{press. at } B = g\rho BB';$$

but the pressures at A' and B' are equal (2·2), therefore

$$\text{press. at } B - \text{press. at } A = g\rho \, (AA' - BB')$$

$$\propto \text{ difference of depths.}$$

It follows of course that if A and B are at the same level the pressures at these points are the same, so that this constitutes an extension of the theorem 2·2 to the case in which the straight line AB does not lie wholly in the fluid provided that the fluid is homogeneous.

2·31. If in 2·3 (1) we put $AB = z$, we have

$$p - p' = g\rho z \quad \text{......................(1)}.$$

The same result clearly holds good if the fluid is not homogeneous but ρ denotes the mean density of the fluid in a thin cylinder about the line AB. Also if the fluid between A and B consists of horizontal strata of uniform densities and ρ denotes the density of the stratum of vertical thickness z, then

$$p - p' = \Sigma g\rho z \quad \text{......................(2)},$$

where the summation extends to all the strata which lie between A and B.

And finally, if the density varies continuously and ρ is the mean density of a thin stratum of thickness dz, then

$$p - p' = \int g\rho \, dz \quad \text{......................(3)},$$

where the integration extends along the line between A and B.

2·32. *In a fluid at rest under gravity horizontal planes are surfaces of equal density.*

Let A, B be two points in the fluid in the same horizontal line and A', B' two points at a short distance vertically above A, B and such that $AA' = BB'$.

Let ρ denote the mean density of the fluid between A and A' and ρ' the mean density of the fluid between B and B', so that, using p_A to denote the pressure at A, we have

$$p_A - p_{A'} = g\rho AA'$$

and

$$p_B - p_{B'} = g\rho'BB'.$$

But from 2·2 $p_A = p_B$ and $p_{A'} = p_{B'}$, and since $AA' = BB'$ therefore $\rho = \rho'$. By diminishing the distances AA' and BB' it follows that the density at $A =$ the density at B. Hence the density at all points in the same horizontal plane is the same.

Thus when fluid is at rest under gravity horizontal planes are surfaces of equal pressure (2·2) and equal density. We shall see later that this is a particular case of a more general theorem that when a fluid is at rest in a conservative field of force the surfaces of equal density and the surfaces of equal pressure coincide.

2·33. *When two fluids of different densities at rest under gravity do not mix their surface of separation is a horizontal plane.*

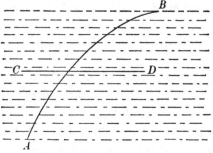

For if the surface of separation AB be other than horizontal it can be intersected by horizontal lines such as CD. At the points C, D the densities will be different, since they lie in the different fluids; and this is in contradiction to the theorem of 2·32.

It follows that the free surface of a liquid at rest is a horizontal plane, and this fact is independent of the shape of the containing vessel, and if the vessel has several branches the free surface stands at the same level in all the branches provided that the density is the same throughout.

2·4. Pressure in heavy homogeneous liquid. The earth's atmosphere produces a pressure which depends upon the force of gravity, the height above the earth's surface and other local conditions. At sea-level it varies about the neighbourhood of 14·5 lb. to the square inch. All surfaces exposed to the atmosphere are subject to this pressure and this applies to liquids with an exposed or 'free' surface.

If in 2·3 we take the point B in the free surface of the liquid, the pressure p' is then the atmospheric pressure, usually denoted by Π. Let the point A be at depth z, then 2·3 (1) becomes

$$p = \Pi + g\rho z \quad\ldots\ldots\ldots\ldots\ldots\ldots(1).$$

If we put $\Pi = g\rho h$, then a horizontal surface at a height h above the free surface of the liquid is called the **effective surface,** and (1) may be written

$$p = g\rho (h + z) \quad\ldots\ldots\ldots\ldots\ldots\ldots(2),$$

so that p varies as the depth below the effective surface.

2·41. Head of Liquid. A pressure due to a *head* of k feet of liquid means a pressure at a depth of k feet below the effective surface.

2·42. Example. *A fine tube of uniform bore is bent into the form of a square and filled with equal volumes of three heavy liquids of densities ρ_1, ρ_2, ρ_3 $(\rho_1 < \rho_2 < \rho_3)$. If the tube is placed with one side vertical, shew that a side of the square will be filled with liquid of the first, third or second kind only according as ρ_2 is $> \frac{1}{3}(2\rho_3 + \rho_1)$ or $< \frac{1}{3}(\rho_3 + 2\rho_1)$ or lies between these values.*   [I.]

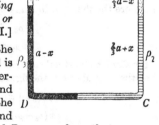

Let a be a side of the square, then the length of tube occupied by each liquid is $\frac{4}{3}a$. First let the liquid ρ_1 fill the uppermost side of the tube and lengths x and $\frac{1}{3}a - x$ of the adjacent sides. Then in the figure, as in 2·2, the pressures at A and B are equal and the pressures at C and D are equal; so that

press. at D − press. at A = press. at C − press. at B,

or $\qquad g\rho_1 x + g\rho_3 (a - x) = g\rho_1 (\frac{1}{3}a - x) + g\rho_2 (\frac{2}{3}a + x)$

or $\qquad x (\rho_3 + \rho_2 - 2\rho_1) = \frac{1}{3}a (3\rho_3 - \rho_1 - 2\rho_2)\ldots\ldots\ldots\ldots(1).$

The necessary and sufficient conditions for this arrangement of liquids to be possible are $0 < x < \frac{1}{3}a$, or from (1)

$$0 < \frac{3\rho_3 - \rho_1 - 2\rho_2}{\rho_3 + \rho_2 - 2\rho_1} < 1.$$

Since $\rho_1 < \rho_2 < \rho_3$ the numerator and denominator are positive, so that the first inequality is satisfied, and the remaining condition is

$$3\rho_3 - \rho_1 - 2\rho_2 < \rho_3 + \rho_2 - 2\rho_1$$

or
$$\rho_2 > \tfrac{1}{3}(2\rho_3 + \rho_1) \quad\quad\quad\quad\quad\dots\dots\dots\dots\dots\dots(2).$$

Secondly let the lowest side CD be filled with the liquid of density ρ_3, and the other liquids be as in the figure. A like argument gives

$$\rho_1(a - x) + \rho_3 x = \rho_2(\tfrac{2}{3}a + x) + \rho_3(\tfrac{1}{3}a - x)$$

or
$$x(2\rho_3 - \rho_1 - \rho_2) = \tfrac{1}{3}a(2\rho_2 + \rho_3 - 3\rho_1) \quad\quad\dots\dots\dots\dots\dots(3)$$

and again the necessary and sufficient conditions are $0 < x < \frac{1}{3}a$, or from (3)

$$0 < \frac{2\rho_2 + \rho_3 - 3\rho_1}{2\rho_3 - \rho_1 - \rho_2} < 1.$$

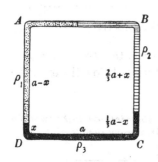 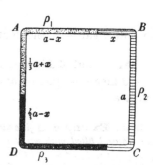

Here the numerator and denominator are positive, since $\rho_1 < \rho_2 < \rho_3$, therefore the first condition is satisfied and the second is

$$2\rho_2 + \rho_3 - 3\rho_1 < 2\rho_3 - \rho_1 - \rho_2$$

or
$$\rho_2 < \tfrac{1}{3}(\rho_3 + 2\rho_1) \quad\quad\quad\quad\quad\dots\dots\dots\dots\dots\dots(4).$$

Finally let a vertical side BC be filled with liquid of density ρ_2, and the other liquids as in the figure. Reasoning as before we have

$$\rho_1(\tfrac{1}{3}a + x) + \rho_3(\tfrac{2}{3}a - x) = \rho_2 a$$

or
$$x(\rho_3 - \rho_1) = \tfrac{1}{3}a(2\rho_3 + \rho_1 - 3\rho_2) \quad\quad\dots\dots\dots\dots(5),$$

and the necessary and sufficient conditions for this arrangement are

$$0 < x < \tfrac{1}{3}a$$

or
$$0 < \frac{2\rho_3 + \rho_1 - 3\rho_2}{\rho_3 - \rho_1} < 1.$$

Now the denominator is positive, hence the first inequality requires that

$$\rho_2 < \tfrac{1}{3}(2\rho_3 + \rho_1) \quad \dots\dots\dots\dots\dots\dots\dots(6)$$

and the second requires that

$$2\rho_3 + \rho_1 - 3\rho_2 < \rho_3 - \rho_1$$

or
$$\rho_2 > \tfrac{1}{3}(\rho_3 + 2\rho_1) \quad \dots\dots\dots\dots\dots\dots\dots(7);$$

and (6) and (7) are together necessary and sufficient for this arrangement.

2·5. Thrust of heavy homogeneous liquid on plane surfaces. *The thrust of a heavy homogeneous liquid on a plane area is equal to the product of the area and the pressure at its centroid.*

Let dS denote a small element of the area S and let z be the depth of dS below the effective surface. Then if ρ is the density of the liquid, since the pressure at depth z is $g\rho z$, therefore the thrust on the element dS is $g\rho z\,dS$ and the total thrust on the area S

$$= \Sigma g\rho z\,dS,$$

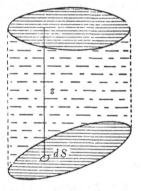

summed for all the elements of area into which S is divided. But

$$\Sigma z\,dS = \bar{z}\Sigma\,dS = \bar{z}S,$$

where $\bar{z}$ is the depth of the centroid of the area S. Therefore the total thrust $= S \times g\rho\bar{z}$ and since $g\rho\bar{z}$ is the pressure at the centroid of the area S, this proves the proposition. But we must note that the thrusts on the elements constitute a set of parallel forces (at right angles to the plane) and we have found the magnitude of the resultant force but *not* the point at which it acts; and the magnitude depends only on the area and the position of its centroid and not on its inclination to the horizontal.

2·51. It is evident from 2·5 that the thrust on the horizontal base of a vessel of the liquid which it contains does not depend on the shape of the vessel but only on the area of the base and the depth of the liquid, being equal to the weight of a cylinder of liquid whose base is that of the vessel and whose height is the depth of the liquid.

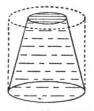

If for example a hollow right circular cone of height h and radius of base r has its vertex uppermost and base horizontal and is just full of liquid of density ρ, the pressure at any point of the base is $g\rho h$ so that the total thrust on the base is $g\rho\pi r^2 h$. But the weight of the liquid is only $\frac{1}{3}g\rho\pi r^2 h$, the volume of the cone being $\frac{1}{3}\pi r^2 h$. Hence the thrust on the base exceeds the weight of the liquid in the cone by $\frac{2}{3}g\rho\pi r^2 h$ and this force must measure the resultant thrust of the liquid on the slant sides of the cone. So that the thrust on the base is made up of the weight of the liquid and the downward reaction of the slant sides on the liquid in the proportion of 1 to 2.

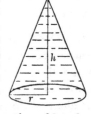

2·511. The reader will note that in the example just considered we used the words 'just full'. It is obvious that the pressure inside a sealed vessel full of liquid can be made to vary considerably by altering its temperature, but we shall use the words 'just full' to signify that there is a point in the liquid at which the pressure vanishes. Thus in the example of **2·51** we assumed that the pressure was zero at the vertex of the cone. The use of the words 'just full' in this sense is a common convention.

2·6. Examples. (i) *A pair of equal vertical lock gates are kept shut by the thrust of water 14 feet deep. Taking the gates as rectangular and plane, each 13 feet wide and supported by vertical hinges 24 feet apart, find the reaction between the gates across the line where they abut.*

Let R be the required reaction, T the thrust of the water on a gate and α the angle which the gates make with the plane through the hinges.

Since the thrust T acts at a point on the vertical bisector of a gate, it is evident by symmetry that the resultant reaction of the hinges must also be a force R. Hence

$$2R \sin \alpha = T$$

or $\qquad\qquad 2R \cdot {}_{13}^{5} = 14 \times 13 \times 7 \times 62 \cdot 5 \, \text{lb. wt.,}$

taking the weight of a cubic foot of water to be 62·5 lb. Whence we find that $R = 46\cdot21$ tons.

(ii) *A parallelogram $ABCD$ is immersed in homogeneous fluid of density ρ, open to the atmosphere (pressure Π), with the side AB in the surface. E is a point in AB such that $AE = \frac{2}{3}AB$. A straight line joins E to a point F in CD. Shew that, if the thrusts on $AEFD$ and $EBCF$ are equal, DF is given by*

$$6DF (3\Pi + 2g\rho b) = AB (6\Pi + 5g\rho b),$$

where b is the depth of CD. [M. T.]

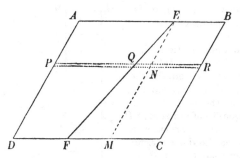

As this problem involves only the thrusts on plane areas, it is a case in which the theorem of 2·5 might be applied. But its application requires a knowledge of the positions of the centroids of the areas $AEFD$ and $EBCF$ and it is simpler to obtain the result by direct integration.

At depth z take a horizontal strip of breadth dz, say PQR in the figure; and let EM parallel to BC meet PR in N and DC in M. Then

$$PQ = PN - QN = AE - \frac{z}{b}FM = \frac{2}{3}AB - \frac{z}{b}(\tfrac{2}{3}AB - DF);$$

and $\qquad QR = AB - PQ = \tfrac{1}{3}AB \left(1 + \frac{2z}{b}\right) - DF \cdot \frac{z}{b}.$

And since the pressure at depth z is $\Pi + g\rho z$, we get for the thrust on $AEFD$

$$\int_0^b (\Pi + g\rho z) \left\{ \tfrac{2}{3}AB \left(1 - \frac{z}{b}\right) + DF \frac{z}{b} \right\} dz$$

$$= \tfrac{1}{6}AB \cdot b (3\Pi + g\rho b) + \tfrac{1}{6}DF \cdot b (3\Pi + 2g\rho b);$$

and for the thrust on $EBCF$

$$\int_0^b (\Pi + g\rho z) \left\{ \tfrac{1}{2} AB \left(1 + \frac{2z}{b}\right) - DF \frac{z}{b} \right\} dz$$
$$= \tfrac{1}{18} AB \cdot b \,(12\Pi + 7g\rho b) - \tfrac{1}{6} DF \cdot b \,(3\Pi + 2g\rho b).$$

By equating the two thrusts we obtain the required result.

(iii) *Two vertical tubes of equal uniform bore are connected at their bases by a horizontal tube of small bore: mercury is first poured into them, and then water is poured down one, until there is equilibrium with the mercury surface at a height h above the common surface. Another communication between the tubes is now opened by means of a fine horizontal tube at a height c above the common surface: prove that the length of the column of water which will flow from one tube to the other is the smaller of σc and* $\tfrac{1}{2}\sigma h$, *where σ is the specific gravity of mercury.*   [I.]

Let GH and CD be the free surfaces of the water and the mercury and EF their common surface before

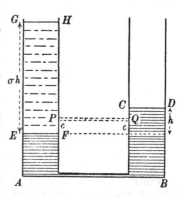

the communication is opened. The height of the column of water GE is then σh.

When a communication PQ is opened and the liquids take up a new equilibrium position, the conditions to be satisfied are equality of pressure at the bottom of the tubes and equality of pressure at P and Q (2·2). And since pressure difference between P and A = pressure difference between Q and B, the mercury columns below the level PQ must be of the same height and likewise the water columns if both liquids extend below that level.

There are two possibilities, $h \leqslant 2c$ or $h > 2c$. In the former case the whole of the mercury will sink below PQ, and to make the pressures at A and B equal, one half the water will pass along PQ to the tube CB, i.e. a column of length $\tfrac{1}{2}\sigma h$. In the latter case the mercury in the tube AH will rise to P and consequently in tube CB it will stand at a height $h - 2c$ above Q. Suppose that in order to balance the pressures at P and Q a column of water of length x is transferred. Then the pressure at Q is due to a column of mercury of length $h - 2c$ and a column of water of length x, while the pressure at P is due to a column of water of length $\sigma h - x$. Hence

$$\sigma(h - 2c) + x = \sigma h - x,$$

so that $$x = \sigma c.$$

It follows that the length of the column of water transferred is σc or $\tfrac{1}{2}\sigma h$ according as $2c$ is less than or greater than h.

EXAMPLES

1. Two cisterns containing water are connected by a pipe. The upper cistern is open, and the free surface of the water in it is 20 feet above the top of the lower cistern. The latter is 4 feet long by 6 feet broad by 5 feet deep, and full of water. Taking the height of the water barometer as 30 feet, calculate the resultant fluid pressures on the top, sides, and base of the lower cistern in tons weight. [36 cubic feet of water weigh 1 ton.]

2. The flat bottom of a vessel is movable, and has an area of 9 square inches. The vessel contains mercury to a height of 3 inches, and above the mercury water to an additional height of 16 inches. Calculate the upward force necessary to keep the bottom in position, assuming that a cubic foot of water weighs 1000 oz., and that the specific gravity of mercury is 13·6. [C.]

3. The lower ends of two vertical tubes of diameters 1 inch and 0·4 inch respectively are connected by a tube. The tubes contain mercury of specific gravity 13·6. If 28 cubic inches of water are poured into the larger tube, by how much is the level of the mercury in the smaller tube raised?

4. Equal volumes of three fluids of different densities, which do not mix, together completely fill a circular tube which is kept in a vertical plane. Prove that, if the densities of the fluids are in arithmetical progression, the common surface of the lightest and heaviest fluids is at an extremity of a horizontal diameter of the circle.

5. A circular tube centre O is filled with three fluids of densities ρ_1, ρ_2, ρ_3 (in descending order of magnitude) and placed in a vertical plane. If $2\alpha, 2\beta, 2\gamma$ be the angles subtended at the centre by the fluids and P be the point on the circumference midway between the ends of the lightest fluid, then the angle θ which OP makes with the vertical is given by

$$\frac{(\rho_2 - \rho_3)}{(\rho_1 - \rho_3)} = \frac{\sin \alpha}{\sin (\alpha + \theta)} \cdot \frac{\sin (\beta - \theta)}{\sin \beta}. \qquad [I.]$$

6. A closed tube in the form of an equilateral triangle contains equal volumes of three liquids which do not mix, and is placed with its lowest side horizontal. Prove that, if the densities of the liquids are in arithmetical progression, their surface of separation will be at points of trisection of the sides of the triangle. [M. T.]

7. A fine glass tube in the shape of an equilateral triangle is filled with equal volumes of three liquids which do not mix, whose densities are in arithmetical progression. The tube is held in a vertical plane, and the side that contains portions of the heaviest and lightest liquids

makes an angle θ with the vertical. Shew that the surfaces of separation divide the sides in the ratio

$$\cos\left(\frac{\pi}{6}-\theta\right):\cos\left(\frac{\pi}{6}+\theta\right). \qquad \text{[I.]}$$

8. A uniform cycloidal tube contains equal volumes of four fluids of densities 1, 2, 3, 4 arranged in the order given: shew that if the axis of symmetry be inclined at an angle $\cot^{-1}\pi$ to the vertical and the tube be completely filled, no fluid will run out. [C.]

9. ABC is a triangular lamina with the side AB in the surface of heavy homogeneous liquid. A point D is taken in AC such that the thrusts on the areas ABD, DBC are equal. Prove that

$$AD:AC=1:\sqrt{2}.$$

10. Find the thrust on a vertical quadrilateral which has one side of length a in the surface, and the opposite side of length b parallel to it at depth h.

If the fluid consists of a top layer of density ρ and thickness $h/2$, and the rest of density σ, prove that the thrust on the quadrilateral is

$$\frac{gh^2}{48}\{(7a+11b)\rho+(a+5b)\sigma\}. \qquad \text{[I.]}$$

11. A rectangular area is immersed in a heavy liquid with two sides horizontal, and is divided by horizontal lines into strips on which the total thrusts are equal. Prove that, if a, b, c are the breadths of three consecutive strips,

$$a(a+b)(b-c)=c(b+c)(a-b). \qquad \text{[M. T.]}$$

12. Equal volumes of two liquids of densities ρ and 3ρ, which do not mix, together just fill a cone which is held with its axis vertical and its vertex uppermost. Prove that the pressure at any point of the base is $3-\sqrt[3]{4}$ times the pressure at the same point when the cone is filled with the lighter liquid.

13. Two lock gates, each 16 feet wide, are closed across a canal, so that each makes an angle of 30° with the line joining their hinges. The depth of water on one side is 12 feet and on the other 6 feet. Find the force with which the two gates press on one another, taking 62·5 lb. as the weight of a cubic foot of water. [I.]

14. The two arms of a U-tube are close together. In the one arm there is water and in the other mercury, so that their common surface is at the lowest point. One quarter of the water is taken out and is poured into the other arm over the mercury. Prove that in the new equilibrium position the difference of heights of the upper surfaces is one-half of what it was formerly. [I.]

15. A U-tube, of uniform section, open at both ends, contains 27 inches of water and such a quantity of mercury that the water is entirely in one of the straight portions of the tube. The straight portions are connected by a fine horizontal tube which passes between them half

an inch above the surface of separation of the water and mercury, the passage through which is originally closed. If this passage is now opened, find in what direction the flow takes place through it, and the volume of liquid which passes before equilibrium is again established. [Take specific gravity of mercury to be 13·5.]   [M. T.]

16. Two vessels each contain oil, water and mercury, and the oil in one is in communication with that in the other by means of a fine horizontal tube and likewise the mercury. Prove that if h_1, h_2 are the heights above a given horizontal plane of the oil-water and water-mercury surface in the one vessel, and h_1', h_2' the corresponding heights in the other, then $(h_1 - h_1')(1 - \rho) = (h_2' - h_2)(\sigma - 1)$, where $\rho =$ specific gravity of oil, and $\sigma =$ specific gravity of mercury.   [I.]

17. A vertical circular cylinder of height $2h$ and radius r, closed at the top, is just filled by equal volumes of two liquids of densities ρ and σ. Shew that, if the axis be gradually inclined to the vertical, the pressure at the lowest point of the base will never exceed

$$g(\rho + \sigma)(r^2 + h^2)^{\frac{1}{2}}.$$ [I.]

ANSWERS

1. 33⅓ tons, 29⅙ tons, 36⅔ tons and 43¾ tons.   2. 18$\frac{47}{68}$ lb.
3. 2·26 inch.   10. $\frac{1}{6}g\rho h^2(a + 2b)$.   13. 24·1 tons.
15. 6·75 inches of water.

Chapter III

CENTRES OF PRESSURE

3·1. Centres of pressure. The resultant thrust of a fluid on a curved surface with which it is in contact in general reduces to a force and a couple—not to a single force. But if the surface be plane the thrusts on all its elements of area constitute a system of parallel forces acting at right angles to the plane, and the point in the plane at which the resultant of these parallel forces acts is called the **centre of pressure** of the area.

3·11. In what follows it is to be assumed, unless the contrary is stated, that the plane area under consideration is immersed in heavy homogeneous liquid, so that in the notation of 2·5 the thrust on an element of area dS at a depth z below the effective surface is $g\rho z\,dS$. The thrusts on all the elements have a resultant $g\rho\bar{z}S$, where $\bar{z}$ is the depth of the centroid and the problem is to find the point in the plane at which this resultant acts.

In the first place we observe that *if the plane of the area be turned about its line of intersection with the effective surface the position on the area of its centre of pressure remains unaltered,*

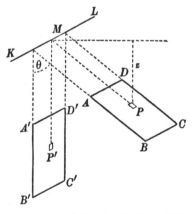

because the effect of the rotation is merely to alter the depths of all points of the area in the same ratio, so that the thrusts on all the elements of area are altered in the same ratio and the relative position of their resultant is therefore unchanged.

For example if $ABCD$ be a plane area inclined at an angle θ to the vertical and P the position of an element dS at depth z, then when the plane is rotated about the line KL, in which it meets the effective surface, until it becomes vertical,

A', B', C', D' and P' being the new positions of A, B, C, D and P, the new depth of dS in the figure is $P'M$ which is equal to PM or $z\sec\theta$, so that the effect of the rotation is merely to multiply the thrust on each element of area by $\sec\theta$ and this will leave the relative position of the resultant thrust unaltered.

Hence in any particular case we lose nothing in generality by supposing the plane of the area to be vertical.

3·12. Formula for the depth of the centre of the pressure of a plane area. Let the plane of the area be vertical and meet the effective surface in the line KL. Let z be the depth of an element dS of the area, h the depth of the centroid G and ζ the depth of the centre of pressure.

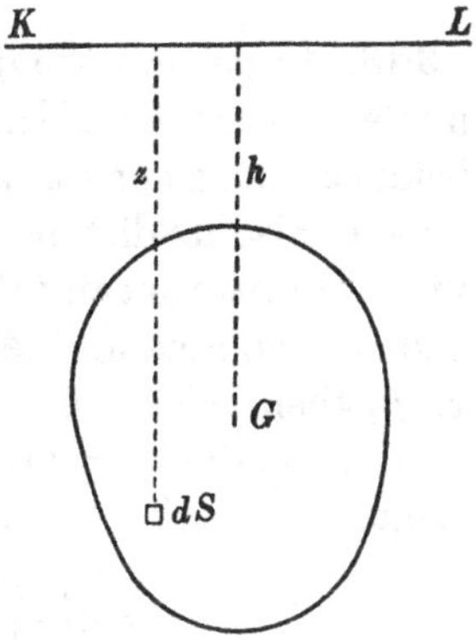

The centre of pressure is the centre of a set of parallel forces of the type $g\rho z\,dS$, and, by taking moments about the line KL, we get for the depth of the centre of pressure

$$\zeta = \frac{\Sigma g\rho z^2\,dS}{\Sigma g\rho z\,dS}$$

or, since g and ρ are constant factors,

$$\zeta = \frac{\Sigma z^2\,dS}{\Sigma z\,dS} \quad\quad\ldots\ldots\ldots\ldots\ldots\ldots(1).$$

We know from the theory of the centre of gravity that the denominator or 'first moment' of the area about KL is equal to Sh, where S denotes the whole area. The numerator or 'second moment' of the area about KL is called the 'moment of inertia'[*] of the area about KL, and if it is denoted by Sk^2, k is called the radius of gyration of the area about KL, and (1) becomes

$$\zeta = k^2/h \quad\ldots\ldots\ldots\ldots\ldots\ldots(2).$$

By the theorem of parallel axes,[†] if k' is the radius of gyration about an axis through G parallel to KL, we have

$$k^2 = k'^2 + h^2,$$

---

[*] See *Dynamics*, Pt I, Ch. XIII.   [†] *Dynamics*, Pt I, p. 188.

so that $$\zeta = \frac{k'^2}{h} + h \quad \dots\dots\dots\dots\dots(3);$$

and the depth of the centre of pressure below the horizontal through G is k'^2/h.

From 3·11 the relative position of the centre of pressure is unaltered by rotating the figure about KL, so that the results obtained above also hold good for any oblique position of the plane provided that distances which we have called 'depths' are taken to be distances from KL, i.e. measured along lines of greatest slope on the plane.

3·13. In general the determination of a centre of pressure involves integration. In 3·12 we only shewed how to find its distance from a horizontal axis and its complete determination involves also its distance from a second axis in the plane. In general terms we can take any set of rectangular axes in the plane of the area and let p denote the pressure at the point (x, y), then using a small rectangular element of area $dx\,dy$, the coordinates ξ, η of the centre of pressure are given by formulae involving double integration, viz.

$$\xi = \frac{\iint xp\,dx\,dy}{\iint p\,dx\,dy}, \quad \eta = \frac{\iint yp\,dx\,dy}{\iint p\,dx\,dy}.$$

It is sometimes convenient to use polar coordinates.

3·2. There are elementary methods by which the position of the centre of pressure can be found in simple cases.

Suppose that the figure $ABCD$ is the area in question in a vertical plane meeting the effective surface in the line KL. At every point of the area erect an ordinate equal to the distance of the point from KL. The ends of these ordinates lie on a plane $A'B'C'D'$ inclined at 45° to the vertical. Then $ABCD$ is the base of a cylindrical or prism-shaped solid cut off by the plane $A'B'C'D'$, and if dS is an element of $ABCD$ at a distance z from KL the volume of the solid standing on dS is $z\,dS$, i.e. proportional to the thrust on dS in the hydrostatic problem. Hence, taking moments about KL to find the position of the resultant thrust, must give the same result as taking the moment about KL of the volume of the solid $ABCDA'B'C'D'$;

and this determines the depth below KL of the centre of gravity of this solid. Consequently the centre of gravity of this solid projects horizontally on to the plane $ABCD$ into the centre of pressure.

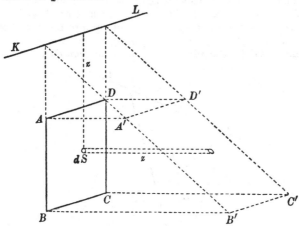

3·21. Examples. (i) *A rectangle with a side in the effective surface.*

Applying the method of 3·2 the resulting solid is a wedge whose centre of gravity G is $\frac{2}{3}$ of the way down the median of its central vertical section, and this point projects horizontally into a point H $\frac{2}{3}$ of the way down the vertical bisector of the rectangle, which is therefore its centre of pressure.

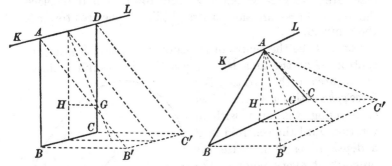

(ii) *A triangle with its vertex in the effective surface and its base horizontal.*

In this case with the vertex A of the triangle in the effective surface the solid is a pyramid on a rectangular base $BCC'B'$. Its centre of gravity G is $\frac{3}{4}$ of the way from A to the centre of the base,* and projects

* *Statics*, p. 164.

into a point H which is $\frac{3}{4}$ of the way down the median from A to the middle point of BC; so that this is the position of the centre of pressure.

(iii) *A triangle ABC with a side AB in the effective surface.*

In this case the solid is a tetrahedron $ABCC'$. If CD is a median of the triangle ABC and E the centroid of the triangle, so that $CE = \frac{2}{3}CD$, then the centre of gravity of the tetrahedron is at G on $C'E$ such that $C'G = \frac{3}{4}C'E$,* and G projects horizontally into a point H on CD such that $CH = \frac{3}{4}CE = \frac{1}{2}CD$. Consequently the centre of pressure of the triangle ABC is the middle point of the median CD.

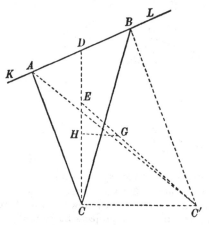

3·3. Effect of additional depth. The effect of lowering a plane area to a greater depth, without rotation, is to displace the centre of pressure along a straight line in the area towards the centroid.

For let C be the centre of pressure of the area S when the centroid G of the area is at depth h. The resultant thrust on the area in this position is a force $g\rho hS$ acting at C. The effect of lowering the area without rotation until the centroid is at a depth h' is to increase the pressure at every point of the area by the same amount

$$g\rho\,(h'-h).$$

Since this extra pressure is uni-

* *Statics*, p. 163.

form over the area, it will have a resultant $g\rho\,(h'-h)\,S$ acting at G (in the new position). Hence in the new position the resultant thrust is compounded of the parallel forces $g\rho hS$ at C and $g\rho\,(h'-h)\,S$ at G, and these give a resultant $g\rho h'S$ acting at a point C' on GC, such that by moments about G

$$GC' : GC = h : h'.$$

3·31. Example. *A plane triangular area is immersed in liquid of uniform density with its plane vertical, one side horizontal and the opposite corner downwards. Its vertical altitude is h, and the horizontal side is at a depth h below the effective surface. Shew that its centre of pressure is at a depth $\frac{11}{8}h$ below the surface.*   [M. T.]

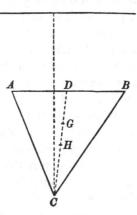

Let S be the area of the triangle ABC, AB the horizontal side and G the centroid. If the side AB were in the effective surface, the thrust on the triangle would be $g\rho \cdot \tfrac{1}{3}h \cdot S$ acting at H, the middle point of the median CD (**2·5** and **3·21** (iii)).

The effect of lowering the triangle in the liquid until AB is at depth h is to add on another thrust $g\rho hS$ at G, so that the total thrust is now $\tfrac{4}{3}g\rho hS$ acting at a point H' in GH such that

$$\tfrac{4}{3}h \cdot GH' = \tfrac{1}{3}h \cdot GH$$

or $$GH' = \tfrac{1}{4}GH.$$

But the depth of G is $\tfrac{4}{3}h$ and that of H is $\tfrac{3}{2}h$, therefore the depth of $H' = (\tfrac{4}{3} + \tfrac{1}{24})\,h = \tfrac{11}{8}h.$

Alternatively we may proceed directly by integration. Thus let PQ be a narrow horizontal strip of the area at depth x and of vertical breadth dx. Let $AB = c$. Then by similar triangles

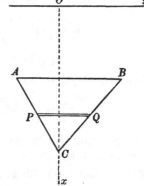

$$PQ : AB = 2h - x : h,$$

since $2h$ is the depth of C;

or $$PQ = c\,(2h - x)/h.$$

Since the pressure at depth x is $g\rho x$, the thrust on the narrow strip is

$$g\rho x \frac{c}{h} (2h - x)\, dx;$$

and the depth ξ of the centre of pressure is given by

$$\xi = \frac{\int_h^{2h} g\rho x^2 \frac{c}{h}(2h-x)\,dx}{\int_h^{2h} g\rho x \frac{c}{h}(2h-x)\,dx} = \frac{\int_h^{2h}(2hx^2-x^3)\,dx}{\int_h^{2h}(2hx-x^2)\,dx},$$

which reduces to $\xi = \frac{11}{8}h$.

3·32. Centre of pressure of a triangular area whose angular points are at depths α, β, γ. The most direct method of obtaining the formula for the depth of the centre of pressure is to assume the triangle to be in a vertical plane and use the expression k^2/h of **3·12**, where h is the depth of the centroid—in this case $\frac{1}{3}(\alpha+\beta+\gamma)$—and k is the radius of gyration of the triangle about the line of intersection of its plane with the effective surface.

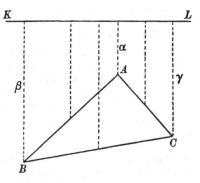

It can be shewn that a uniform triangular lamina of mass M is equimomental with three particles of mass $\frac{1}{3}M$ placed at the middle points of its sides; i.e. it has the same centre of gravity and the same moment of inertia about any line in its plane.

Hence the k^2 for the triangle of area S is the same as for particles $\frac{1}{3}S$ at the middle point of each side. But the depths of these middle points are $\frac{1}{2}(\beta+\gamma)$, $\frac{1}{2}(\gamma+\alpha)$, $\frac{1}{2}(\alpha+\beta)$, so that

$$k^2 = \frac{1}{3}\left\{ \left(\frac{\beta+\gamma}{2}\right)^2 + \left(\frac{\gamma+\alpha}{2}\right)^2 + \left(\frac{\alpha+\beta}{2}\right)^2 \right\}$$

$$= \tfrac{1}{6}(\alpha^2+\beta^2+\gamma^2+\beta\gamma+\gamma\alpha+\alpha\beta).$$

Hence the depth of the centre of pressure is given by

$$\frac{k^2}{h} = \frac{1}{2}\frac{\alpha^2+\beta^2+\gamma^2+\beta\gamma+\gamma\alpha+\alpha\beta}{\alpha+\beta+\gamma}.$$

By rotating the plane of the triangle about its line of intersections with the effective surface it is then easy to shew that the same formula holds good for a triangle in any position, for the depths in the new position all bear the same ratio to what were the depths in the vertical position. It is also evident that the thrust on a triangular area in any position is equivalent to three forces acting on the middle points of the sides and proportional respectively to the depths of those points.

3·321. The equimomental property quoted in **3·32** can be demonstrated thus. Let β, γ be the distances of the points B, C of a triangle ABC from any line AX through A in the plane of the triangle. Let BC meet this line in D.

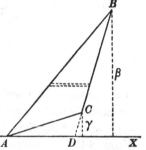

By taking narrow strips of the triangle ABD parallel to AX, it is easy to see that, if y is the distance of a strip from AX and dy its breadth, its area is

$$AD\left(1-\frac{y}{\beta}\right)dy,$$

so that the moment of inertia about AX of the area ABD is

$$\frac{AD}{\beta}\int_0^\beta y^2\,(\beta-y)\,dy = \tfrac{1}{12}AD\,.\,\beta^3.$$

Similarly the moment of inertia of ACD is $\tfrac{1}{12}AD\,.\,\gamma^3$.

Therefore the moment of inertia of ABC is $\tfrac{1}{12}AD\,(\beta^3-\gamma^3)$, but its area is $S=\tfrac{1}{2}AD\,(\beta-\gamma)$, therefore its moment of inertia is

$$\tfrac{1}{6}S\,(\beta^2+\beta\gamma+\gamma^2)$$

or

$$\tfrac{1}{3}S\left\{\left(\frac{\beta+\gamma}{2}\right)^2+\left(\frac{\beta}{2}\right)^2+\left(\frac{\gamma}{2}\right)^2\right\};$$

and this is also the moment of inertia about AX of three particles $\tfrac{1}{3}S$ at the middle points of the sides of the triangle ABC. This establishes the equality of the moments of inertia about AX and it is easy then to prove that the two systems have the same centre of gravity and the same moments of inertia about a line parallel to AX through the centre of gravity and by the theorem of parallel axes about any parallel line. But the direction of AX is arbitrary so the theorem is proved.

3·33. Example. *The centre of pressure of a parallelogram completely immersed is in one of its diagonals. Prove that the other diagonal is horizontal.*   [M. T.]

Let $ABCD$ be the parallelogram, G its centre and P, Q, R, S the middle points of its sides; let h be the depth of G, and p, q, r, s the depths of P, Q, R, S; and let the centre of pressure be in the diagonal AC.

The triangles ABC and ADC are of equal area, so the thrusts on them can be represented by forces kp, kq, kh acting at P, Q, G and

kr, ks, kh acting at R, S, G, and these six forces have a resultant acting at a point on AC. But the distances of the lines of action of the four forces kp, kq, kr, ks from AC are the same, so we must have $p+q=r+s$. Also the middle points of PQ and SR both lie on BD, and their depths being $\frac{1}{2}(p+q)$ and $\frac{1}{2}(r+s)$ are equal, so that BD must be horizontal.

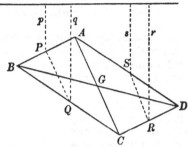

3·4. Centre of pressure of a circular area. Let the circle be of radius a with its centre G at a depth h below the effective surface and let its plane be vertical.

From 3·12 the centre of pressure is at a depth k'^2/h below the centre of the circle, where k' is the radius of gyration of the circle about a horizontal through G in its plane, i.e. about a diameter. But $k'^2=a^2/4$,* therefore the centre of pressure is at a depth $a^2/4h$ below G.

This result may be obtained directly by integration or by the following considerations: let the circular area be the base of a hemisphere. Then the liquid in this hemisphere is kept at rest by the thrusts on its bounding surfaces. The thrusts on the curved surface all pass through the centre G, therefore the thrust on the plane face must have a moment about G equal and opposite

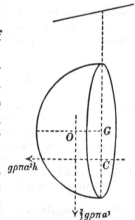

to that of the weight of the liquid in the hemisphere. But the weight is $\frac{2}{3}g\rho\pi a^3$ acting vertically through the centre of gravity O of the hemisphere, which is such that $GO=\frac{3}{8}a$;†

* *Dynamics*, Pt I, p. 190.
† *Statics*, p. 172.

and the thrust on the vertical circle is $g\rho\pi a^2 h$ acting through the centre of pressure C.

Therefore by moments about G

$$g\rho\pi a^2 h \cdot GC = \tfrac{2}{3}g\rho\pi a^3 \cdot \tfrac{3}{8}a$$

or $$GC = a^2/4h.$$

When the circular area is inclined to the vertical it is clear from **3·11** that the same formula will represent the distance GC in the plane of the circle, if h denotes the distance of G from the effective surface measured along the plane of the circle.

3·41. Centre of pressure of an elliptic area when its major axis is vertical or along a line of greatest slope. Let the plane of the circle of **3·4** meet the effective surface in the line KL. Take any other plane through KL and project the circle on to it by drawing horizontal lines at right angles to the plane of the circle; thus forming a circular cylinder with a circular section and an oblique (elliptic) section. If we consider the equilibrium of the liquid in this cylinder and resolve in the direction of its generators, we see that the thrust on the circle acting through its centre of pressure C must be balanced by the horizontal component of the thrust on the ellipse, so that the resultant thrust on the ellipse

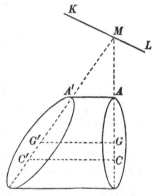

must act at the point C' in which the horizontal through C meets the ellipse. Thus in the figure the points A, G, C are the orthogonal projections of A', G', C'. Hence we have

$$\frac{G'C'}{A'G'} = \frac{GC}{AG} = \frac{1}{4}\frac{AG}{MG} \quad (3\cdot4)$$

$$= \frac{1}{4}\frac{A'G'}{MG'};$$

or $G'C' = \tfrac{1}{4}A'G'^2/MG'$, where M is in the effective surface; and the result is clearly independent of the inclination of the ellipse to the vertical.

3·5. *Position of the centre of pressure referred to coordinate axes through the centroid of the area.*

Let the plane of the area be vertical. Let h be the depth of its centroid G. Take rectangular axes Gx, Gy and let

Gx make an angle θ with the horizontal, lying below the horizontal through G as in the figure.

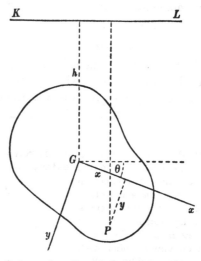

The depth of the point P whose coordinates are x, y is then $h + x \sin \theta + y \cos \theta$, so that the thrust on a small element of area dS at P is

$$g\rho \, (h + x \sin \theta + y \cos \theta)\, dS.$$

Hence, by moments about the axes, the coordinates ξ, η of the centre of pressure are given by

$$\xi = \frac{\Sigma x \, (h + x \sin \theta + y \cos \theta)\, dS}{\Sigma \, (h + x \sin \theta + y \cos \theta)\, dS}, \qquad \eta = \frac{\Sigma y \, (h + x \sin \theta + y \cos \theta)\, dS}{\Sigma \, (h + x \sin \theta + y \cos \theta)\, dS}$$
$$\dots\dots(1).$$

In the summations $\sin \theta$ and $\cos \theta$ are definite constants; also $\Sigma x \, dS$ and $\Sigma y \, dS$ are both zero because the origin is the centroid of the area.

$\Sigma x^2 \, dS$ is the moment of inertia of the area
about $Gy = A$ say,

$\Sigma y^2 \, dS$ is the moment of inertia of the area
about $Gx = B$ say,

and $\qquad \Sigma xy \, dS$ is the product of inertia of the area
with regard to the axes and $= F$ say.

Thus the expressions for the coordinates reduce to

$$\xi = \frac{A \sin\theta + F \cos\theta}{hS}, \quad \eta = \frac{F \sin\theta + B \cos\theta}{hS} \quad ...(2),$$

where S denotes the whole area.

If we eliminate θ we get an equation of the second degree shewing that if the area turns round G in its own plane the locus of the centre of pressure on the area is a conic.

If we suppose the axes chosen to be the principal axes at G,[*] then the product of inertia F vanishes and we may put $A = Sa^2$ and $B = Sb^2$, where a, b are the principal radii of gyration of the area, and then (2) become

$$\xi = \frac{a^2 \sin\theta}{h}, \quad \eta = \frac{b^2 \cos\theta}{h} \quad ....(3),$$

and the locus of the centre of pressure as θ varies is

$$\frac{\xi^2}{a^4} + \frac{\eta^2}{b^4} = \frac{1}{h^2} \quad ....(4).$$

Again, the ellipse $\dfrac{x^2}{a^2} + \dfrac{y^2}{b^2} = 1$ $\quad ....(5),$

where a, b have the same meanings as above, is called the principal momental ellipse of the area at G, or its central ellipse. The polar of the point (ξ, η) with regard to this ellipse is

$$\frac{x\xi}{a^2} + \frac{y\eta}{b^2} = 1;$$

or, substituting from (3),

$$x \sin\theta + y \cos\theta = h \quad ....(6).$$

But this is the equation of a straight line parallel to the surface and at a depth h below G.

It follows that the *centre of pressure is the pole of this line with regard to the central ellipse.*

Cor. Since the central ellipse of a square is a circle, therefore, when a square is completely immersed with its plane vertical, its centre of pressure lies vertically below its centre, no matter what are the inclinations of its sides to the vertical.

---

[*] *Dynamics*, Pt I, p. 193.

3·51. Example. *An elliptic disc is just immersed in homogeneous liquid with its plane vertical. Find the locus on the disc of its centre of pressure.*

Take the axes of the ellipse as co-ordinate axes with Gx making an angle θ with the horizontal as in 3·5. If a, b are semiaxes of the ellipse and S its area, its principal moments of inertia are $\frac{1}{4}Sa^2$, $\frac{1}{4}Sb^2$, so that from 3·5 (3) the coordinates of the centre of pressure are

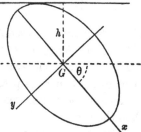

$$\xi=\frac{a^2\sin\theta}{4h}, \quad \eta=\frac{b^2\cos\theta}{4h} \quad \text{...(1).}$$

But in this case, since the ellipse is just immersed, h is the central perpendicular on a tangent to the ellipse; so that

$$h^2=a^2\sin^2\theta+b^2\cos^2\theta \quad\text{.....................(2)}$$

and from (1) and (2) we get

$$\frac{\xi^2}{a^2}+\frac{\eta^2}{b^2}=\frac{1}{16}$$

as the required locus, viz. an ellipse similar and similarly situated to the given ellipse.

EXAMPLES

1. A rectangular door in the vertical side of a reservoir can turn freely about its lowest edge, and is fastened at its two upper corners. The door is 3 feet wide and 6 feet high, and its upper edge is 5 feet below the water-level. Determine the reactions at the upper corners, assuming them to be equal, and taking a cubic foot of water to weigh 62·5 lb.

2. A ditch is 5 feet wide at the top, 3 feet at the bottom and 4 feet deep: the ditch is closed at one end by a board, supported at its four corners; calculate the pressures on the supports when the ditch is full of water on one side of the board, taking a cubic foot of water to weigh 62·5 lb.   [C.]

3. In the vertical side of a vessel containing water there is a square trap door, opening freely outwards about a hinge in its upper edge, two sides of the square being horizontal. The length of the side is 3 cm., and the depth of the hinge below the surface of the water is 9 cm. Find the least force (in grams) that will keep the trap door closed.   [C.]

4. Calculate the resultant pressure (in tons weight) on a vertical quadrilateral dam $ABCD$ in which AD, BC are horizontal, and AB, CD make an angle of 45° with the horizontal; AD is uppermost, the length of AD is 100 feet and that of BC is 50 feet. Find also the position of the centre of pressure.   [C.]

5. Shew that, if any inclined plane surface wholly immersed in liquid be lowered, always remaining parallel to its former position, with its centre of gravity moving along a vertical straight line, the locus in space of the centre of pressure is a hyperbola. [I.]

6. When the depth of the liquid is increased by an amount a the depth of the centre of pressure is found to increase by y, and when, instead, the depth of the liquid is increased by b that of the centre of pressure is found to increase by z. Shew that the depth of the centre of gravity of the area in the original state of the liquid is

$$ab(b-a+y-z)/(az-by).$$

7. A rectangular sluice-gate is placed in the vertical wall of a tank and is free to swing, so that its top edge moves outwards on a horizontal axis, which is distant a from the top edge and b from the bottom edge of the sluice. Shew (i) that, however great the depth of the water in the tank, the sluice will not open unless a be greater than b, and (ii) that the sluice will open before it is completely immersed if a be greater than $2b$. [M. T.]

8. If a square of side $2a$ is completely immersed in homogeneous liquid in a vertical plane with its centre at depth d, prove that the centre of pressure is vertically below the centre of the square and at distance $a^2/3d$ from it, whatever be the inclination of the sides of the square to the vertical. [M. T.]

9. Find the centre of pressure of a regular hexagon of side a with one side in the surface. [M. T.]

10. One wall of a tank slopes inwards from the bottom at an angle θ to the vertical and contains a triangular trap-door, of weight W, which is hinged about a horizontal side BC, has the vertex A lower than BC, and can open outwards. The vertical heights of the vertices above the bottom of the tank are a, b, b. Prove that, if water be poured into the tank to a height h so that the trap-door is entirely below the surface, it will remain closed provided that

$$h < \frac{W}{\Delta s}\sin\theta + \tfrac{1}{3}(a+b),$$

Δ being the area of the triangle and s the weight of unit volume of water. [M. T.]

11. A parallelogram has the highest angular point in the surface of a liquid and one diagonal horizontal. Shew that the depth of its centre of pressure is $\frac{7}{12}$ of the depth of the lowest point. [I.]

12. A parallelogram, whose plane is vertical and centre at a depth h below the surface, is totally immersed. Shew that if a and b are the lengths of the projections of its sides on a vertical line, then the depth of its centre of pressure will be $h + \dfrac{a^2 + b^2}{12h}$. [I.]

13. A vessel contains three fluids of densities ρ, 2ρ and 3ρ. A triangle of area A is supported with one side in the surface of the fluid of density ρ and the opposite vertex in the fluid of density 3ρ. If $3h$, $2h$ and h are the depths of this vertex below the upper surface of the three fluids, prove that, neglecting atmospheric pressure, the thrust on each face of the triangle is $\frac{5}{3}g\rho Ah$.

Find also the depth of the centre of pressure. [M. T.]

14. A quadrilateral is immersed vertically having two sides of lengths $2a$, a parallel to the surface at depths h, $2h$ respectively. Shew that the depth of the centre of pressure is $3h/2$. [M. T.]

15. Shew that the depth of the centre of pressure of a trapezium, of which one side of length a is in the surface and the parallel side of length b is at a depth h, is $\dfrac{a+3b}{a+2b}\dfrac{h}{2}$, neglecting the pressure of the atmosphere. [M. T.]

16. A closed vessel is in the form of a right pyramid of height a, its base being a hexagon of side $2a$ and its slant faces six equal uniform isosceles triangles each of weight $\dfrac{1}{\sqrt{3}}g\rho a^3$, where ρ is the density of water. One of these triangles is hinged to the base along its lower edge about which it can turn freely. The vessel is placed on a horizontal board with this face closed and water is poured in from a small hole at the apex. Shew that this face remains closed, so long as the height y of the water satisfies the inequality

$$a^4 > 2y^3(2a-y). \qquad \text{[M. T.]}$$

17. A rectangular hole in the vertical wall of a vessel containing water is closed by a plate secured along the top and bottom edges only, these edges being horizontal. The height of the hole from top to bottom is b and the water in the vessel rises just to the top edge of the plate. Shew that if there is no bending moment at the top and bottom edges, the maximum bending moment in the plate is at a distance $b/\sqrt{3}$ below the water level and is of amount $wb^3/9\sqrt{3}$ per unit width of the plate, w being the weight per unit volume of water. [M. T.]

18. The end of a horizontal pipe is closed by a sphere of the same radius a as the internal section of the pipe. The sphere is hinged to the pipe at its highest point. If the pipe is just full of liquid of density ρ, prove that the moment about the hinge of the liquid pressure on the sphere is $g\rho\pi a^4$. [M. T.]

19. A cube, with edges of length $2a$, is immersed in a liquid and has one edge in the surface and two faces through that edge equally inclined to the horizontal. Find the centres of pressure of all the faces. [M. T.]

20. The centre of a cube of edge $2a$ is at such a fixed depth h below the surface of liquid that the cube is completely immersed in any position. For any position of the cube in which two faces remain vertical, P is the centre of pressure of one of the other faces. Shew that the greatest distance of P from the centre of the face is $a^2/3\sqrt{(h^2-a^2)}$. Atmospheric pressure is neglected. [M. T.]

21. A tunnel, of rectangular section, of height h feet, is closed by a heavy uniform metal door, inclined at an angle α to the vertical, and swung on hinges along the roof of the tunnel. Shew that if the door is to open automatically just when the level of water in the tunnel rises to the roof, the weight per square foot of the door must be equal to that of $\frac{2}{3}h$ cosec α cubic feet of water. [M. T.]

22. One end of a horizontal pipe of circular section is closed by a vertical door hinged to the pipe at the top. Find the moment about the hinge of the liquid pressure when the pipe is (1) full, (2) half-full of liquid. [I.]

23. The axis of a cylindrical vessel containing liquid is inclined to the vertical at an angle α. Prove that the distance of the centre of pressure of the base from the centre of the base is $r^2\tan\alpha/4b$, where r is the radius of the base and $\pi r^2 b$ the volume of the liquid. [I.]

24. A circular flap, 2 feet in diameter, is used to close a hole in the side of a tank; it is kept in place by bolts at the highest and lowest points of the flap. Calculate the forces on these bolts when the water is 5 feet above the top of the flap, taking the weight of a cubic foot of water to be 62·5 lb. [C.]

25. A circular lamina of radius 1 foot is totally immersed in water with a horizontal diameter fixed at a depth of 3 feet. Shew that if the lamina be rotated about this diameter, the centre of pressure lies on a vertical circle of diameter 1 inch. [M. T.]

26. A plane lamina consists of a circular disc (radius a and centre O) from which a circular portion (radius $a/2$ and centre P) has been cut. The lamina is completely immersed in a homogeneous fluid with its plane vertical and P vertically below O. If OP is equal to $a/2$ and the centre of pressure of the lamina is at O, prove that O is at a depth $\frac{11}{8}a$ below the surface of the fluid. [M. T.]

27. A hole in the side of a ship is closed by a circular door 5 feet in diameter hinged at the highest point and held inside against the water pressure by a fastening at its lowest point. If the highest and lowest points of the door are at a depth of 4 and 8 feet, shew that the least force exerted by the fastening must be 1·78 tons. [M. T.]

28. A circular disc of radius a is completely immersed with its plane vertical in a homogeneous fluid. If h is the depth of the centre below the free surface of the fluid, prove that the distance between the centres of pressure of the two semi-circles into which the disc is divided by its horizontal diameter is $6\pi a(4h^2-a^2)/(9\pi^2 h^2-16a^2)$. [M. T.]

29. A flat circular plate, of radius a, lies in a plane inclined at 30° to the horizontal, and is subjected to water pressure on one face. The centre of pressure is at a distance $\frac{1}{16}a$ from the geometrical centre. Shew that the geometrical centre is at a depth $2a$ below the free surface of the water. [M. T.]

30. A circular sector centre O and radius a, symmetrical about the radius OP, is completely immersed with P in the surface and OP vertical. Determine the depth of its centre of pressure. [I.]

31. Shew that, if a lamina totally immersed in fluid is a quadrant of a circle of radius a of which the centre is in the surface, the locus of the centre of pressure in the lamina lies on a straight line of length

$$3a\sqrt{2}\,(\pi-2)/16. \qquad \text{[I.]}$$

32. One edge of a regular tetrahedron is in the surface of water, the opposite edge being horizontal. Find in terms of the side the depths of the centres of pressure of the four faces. [C.]

33. A square lamina is wholly immersed in a heavy homogeneous fluid with its plane vertical and one corner in the surface; if it be turned in its own plane about this corner, and is always immersed, shew that the locus of the centre of pressure in the lamina is a straight line. [I.]

ANSWERS

1. 1968·75 lb.   **2.** 333⅓ lb., 583⅓ lb.   **3.** 49·5.

4. 581·28. On the bisector of AD and BC at a depth 15⅔ feet.

9. At a depth $\dfrac{23\sqrt{3}}{36}\,a$.   **13.** $\frac{63}{85}h$.

19. Vertical faces at depth $\dfrac{7a}{3\sqrt{2}}$; upper slant faces ⅓a from centre of face; lower faces ⅓a from centre.

22. (1) $\frac{5}{8}g\rho\pi a^4$;   (2) $g\rho a^4\left(\dfrac{2}{3}+\dfrac{\pi}{8}\right)$, where a is the radius.

24. 564·5 lb., 613·6 lb.

30. $\dfrac{a}{4}\dfrac{15\alpha-16\sin\alpha+3\sin\alpha\cos\alpha}{3\alpha-2\sin\alpha}$, where 2α is the angle.

32. $\dfrac{3a}{4\sqrt{2}}$, $\dfrac{a}{2\sqrt{2}}$, where a is the side.

Chapter IV

THRUSTS ON CURVED SURFACES.
FLOATING BODIES

4·1. The principle of Archimedes. *When a solid is wholly or partially immersed in heavy fluid at rest the resultant thrust of the fluid on the solid is equal and opposite to the weight of fluid displaced by the solid and acts in the vertical line through the centre of gravity of the fluid displaced.*

To prove this theorem it is only necessary to observe that the resultant thrust of the fluid on the solid depends only on the shape of the solid and the position which it occupies and is clearly independent of the substance of which the solid is composed, and that the thrust of the fluid on all bodies which would exactly fill the same space would be the same. If then we imagine the solid to be removed and the space which it occupied to be filled up with fluid of the same kind as the surrounding fluid, this mass, which we may call 'the displaced fluid', would be in equilibrium under the action of its weight and the thrusts of the surrounding fluid upon it. Hence the resultant thrust of the fluid upon any solid which would fill the same space must be a force equal and opposite to the weight of fluid displaced and act upwards through the centre of gravity of the fluid displaced. This force is called the **force of buoyancy**, and the centre of gravity of the fluid displaced is called the **centre of buoyancy.**

4·11. The theorem of **4·1** admits of extension to a solid immersed in a fluid in equilibrium under the action of any given external field of force. The thrust of the fluid on the solid would be the same as the thrust upon anything else which occupies the same space. But if the solid were removed and the gap in the fluid were filled up with fluid having the same law of density as the surrounding fluid, this fluid which fills the gap would be in equilibrium under the action of the external field of force and the thrusts of the surrounding fluid. Therefore the resultant thrust of the surrounding fluid must be equal and opposite to the resultant of the external forces upon the fluid which would fill the gap.

4·2. To find the thrust on any surface exposed to the pressure of heavy fluid. If the surface be curved the resultant thrust may not be a single force, but we indicate how to find the vertical component of thrust and the component in any assigned horizontal direction.

Let the surface on which the thrust is to be calculated be bounded by a curve $PQRS$. From every point of this curve draw vertical lines to meet the effective surface of the fluid in a curve $P'Q'R'S'$. These lines form a cylinder; and the only vertical forces on the fluid in this cylinder are its weight and the vertical component of the reaction of the surface $PQRS$ upon it, and in equilibrium these must be balancing

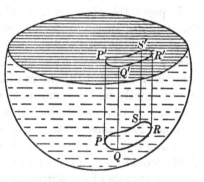

forces. But the reaction of the surface is equal and opposite to the thrust of the fluid upon it. Hence the vertical component of the thrust of the fluid on the surface $PQRS$ is equal to the weight of the fluid in the cylinder

$$PQRSP'Q'R'S'$$

which may be described as 'the superincumbent fluid'.

We note that if the fluid is a liquid with an atmosphere above the surface $P'Q'R'S'$, the vertical component of the thrust on $PQRS$ is the weight of liquid between

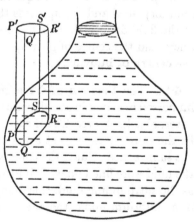

$PQRS$ and $P'Q'R'S'$ together with the atmospheric pressure on $P'Q'R'S'$.

When the fluid presses on the surface $PQRS$ from below as in the second figure, the pressure at each point being due to the depth below the effective surface, it is clear that if we draw vertical lines as before to meet the plane of the effective surface in the curve $P'Q'R'S'$, then the vertical component of the upward thrust of the fluid on $PQRS$ is equal to the weight of fluid which would fill the cylinder $PQRSP'Q'R'S'$.

We have already had an example of this result in 2·51 where it was seen that the upthrust of liquid on the slant sides of a cone full of liquid is $\frac{2}{3}g\rho\pi r^2h$, and this is the weight of liquid which will fill the space between the cone and a circumscribing cylinder.

4·21. Horizontal thrust. To find the component in an assigned horizontal direction of the thrust of heavy fluid on a surface $PQRS$, it is only necessary to draw horizontal lines in the assigned direction through every point of the boundary curve $PQRS$ and take a vertical cross section

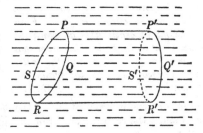

$P'Q'R'S'$

of the cylinder formed by these lines. Then the only horizontal forces acting on the fluid in this cylinder in the direction of its generators are the horizontal component in the assigned direction of thrust on the surface $PQRS$ and the thrust on the plane end $P'Q'R'S'$, and in equilibrium these forces must balance one another. Hence the required component of thrust is equal to the thrust on $P'Q'R'S'$; i.e. the area of this figure multiplied by the pressure at its centroid.

4·22. Whenever a curved surface exposed to the pressure of heavy fluid is bounded by a *plane* curve we can find the thrust on the curved surface in the following way. Consider the equilibrium of the fluid enclosed by the given curved surface S and a plane boundary A. The resultant thrust of the surrounding fluid on S together with its thrust on A must together balance the weight of the fluid enclosed by S and A. Hence by calculating the weight of the enclosed fluid and the thrust on A (its area multiplied by the pressure at its centroid) the thrust on S can be found.

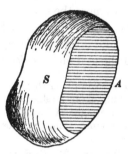

4·23. Examples. (i) *A hemispherical bowl whose mass is* 100 *grams is placed with its rim downwards on a horizontal plane which it fits closely. Water is poured into the bowl through a hole in the curved surface. Find the height in centimetres at which the water must be in the bowl in order that the bowl may be lifted and the water begin to escape between the plane and the bowl.* [M. T.]

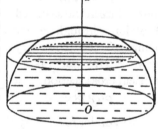

Let a be the internal radius of the bowl and h the required height of the water.

The volume of water required is

$$\int_0^h \pi y^2\,dx = \int_0^h \pi\,(a^2 - x^2)\,dx = \pi\,(a^2 h - \tfrac{1}{3}h^3).$$

But the volume of a cylinder of height h on the same base is $\pi a^2 h$, therefore the volume between the spherical water surface and this cylinder is $\tfrac{1}{3}\pi h^3$, and by 4·2 the upward thrust on the hemisphere is equal to the weight of this volume of water, i.e. $\tfrac{1}{3}\pi h^3$ c.c. of water.

Therefore $\tfrac{1}{3}\pi h^3 = 100$

or $h = (300/\pi)^{\tfrac{1}{3}}.$

By using logarithms we find that $h = 4·57$ cm.

(ii) *A hemispherical surface of radius a is immersed in liquid of density* ρ *with its centre at a depth h and its base inclined at an angle* θ *to the horizontal. Find the resultant thrust on the surface.* [M. T.]

Let X, Y be the horizontal and vertical components of the required thrust on the curved surface.

The thrust on the plane face of the hemisphere is $g\rho\pi a^2 h$ and the weight of liquid it contains is $\frac{2}{3}g\rho\pi a^3$. Since these four forces are in equilibrium, we have by resolving horizontally and vertically

$$X = g\rho\pi a^2 h \sin\theta$$

and
$$Y = g\rho\pi a^2 h \cos\theta + \tfrac{2}{3}g\rho\pi a^3.$$

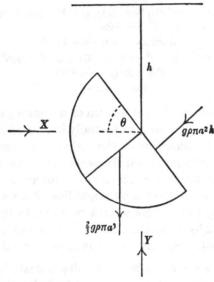

Hence the resultant thrust is

$$(X^2 + Y^2)^{\frac{1}{2}} = g\rho\pi a^2 \{h^2 + \tfrac{4}{9}a^2 + \tfrac{4}{3}ah\cos\theta\}^{\frac{1}{2}}.$$

If the curved surface of the hemisphere be taken to be uppermost the last term in the result will have a minus sign.

(iii) *A portion of a sphere cut off by two planes through its centre inclined to each other at an angle $\pi/4$ is just immersed in a liquid with one face in the surface. Find the resultant thrust on the curved surface and shew that it makes an angle $\tan^{-1}(\pi/2 - 1)$ with the horizontal.* [M. T.]

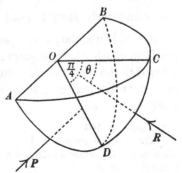

Let ACB, ADB be the plane faces, the former being in the surface of the liquid. Let the resultant thrust R on the curved surface make an angle θ with the horizontal. Let a be the radius of the sphere. The centroid of the semicircle ADB is at a distance $4a/3\pi$

from the centre.* Therefore the thrust P on the plane face ADB is equal to

$$gp \cdot \tfrac{1}{2}\pi a^2 \cdot \frac{4a}{3\pi} \cdot \frac{1}{\sqrt{2}} = \frac{\sqrt{2}}{3} gpa^3.$$

Again the volume of the wedge-shaped solid bears to the volume of the whole sphere the ratio $\dfrac{\pi}{4} : 2\pi$; so that the volume $ABCD$ is $\tfrac{1}{6}\pi a^3$;† and the weight of liquid displaced is $\tfrac{1}{6}gp\pi a^3$. But this force of buoyancy is the resultant of R and P. Therefore

$$R\cos\theta = P\cos(\pi/4) = \tfrac{1}{3}gpa^3$$

and $$R\sin\theta = \tfrac{1}{6}gp\pi a^3 - P\sin(\pi/4) = \tfrac{1}{6}gp\pi a^3 - \tfrac{1}{3}gpa^3;$$

so that $$R = \tfrac{1}{6}gpa^3(\pi^2 - 4\pi + 8)^{\frac{1}{2}}$$

and $$\tan\theta = \pi/2 - 1.$$

4·3. Conditions of equilibrium of a floating body. When a body floats freely wholly or partially immersed in a fluid the only vertical forces acting on the body are its weight and the force of buoyancy (**4·1**), and the necessary and sufficient conditions of equilibrium are that these two forces should be equal and opposite and in the same straight line. From **4·1** it follows that the weight of the body must be equal to the weight of fluid displaced by it and the centres of gravity of the body and of the fluid displaced must be in the same vertical line.

4·31. When a solid of mean density ρ floats in a fluid of density $\rho'(>\rho)$ only the fraction ρ/ρ' of the volume is immersed. For if V denotes the volume of the solid and V' that of the fluid displaced, since their weights are equal,

$$g\rho'V' = g\rho V$$

or $$V' = \frac{\rho}{\rho'}V.$$

The solid clearly cannot float if $\rho > \rho'$.

4·32. When a body floats partly immersed in one fluid and partly in another (e.g. in water and air), the weight of the body is equal to the total weight of fluid displaced, and the centres of gravity of the body and of the whole fluid displaced are in the same vertical line.

---

* *Statics*, p. 169.

† The volume may also be found from the theorem of Pappus (*Statics*, p. 176), since it is traced by the area ACB turning about AB through an angle $\pi/4$.

4·33. Bodies floating under constraint.

(i) *One point fixed.* Let a floating body be free to turn about a fixed point O. Let W, W' be the weights and G, G' the centres of gravity of the body and the fluid displaced. It is clear that the horizontal thrust of the fluid on the body in any

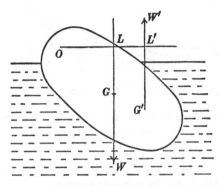

direction is zero. This can be proved by projecting the boundary surface of the part immersed on to a vertical plane perpendicular to an assigned direction as in 4·21, when opposing parts of the surface will be seen to contribute equal and opposite forces in the assigned direction. Alternatively, if the horizontal thrust were not zero a freely floating body would have a horizontal acceleration.

Hence the only forces acting on the body are its weight W acting through G, the force of buoyancy W' acting through G' and the reaction at O. Since W and W' are vertical, the reaction at O must be vertical and the three must be coplanar. Hence it is necessary that the points O, G, G' should lie in a vertical plane, and the moments of W and W' about O must balance; i.e. in the figure

$$W \cdot OL = W' \cdot OL'.$$

(ii) *Two points fixed.* When two points O, O' in the body are fixed, this is equivalent to fixing a line in the body, and the necessary and sufficient condition of equilibrium is that the moments about this line of the weight of the body and the force of buoyancy must be equal and opposite.

4·84. Examples. (i) *A thin uniform wooden rod AB is in equilibrium in an inclined position with one end AC immersed in a bowl of water and one point D supported on the edge of the bowl. Shew that the specific gravity σ of the wood is* $\dfrac{AC}{AB}\cdot\dfrac{2AD-AC}{2AD-AB}$, *and that the greatest fraction of the length of the rod which can remain immersed is* $1-\sqrt{(1-\sigma)}$.   [M. T.]

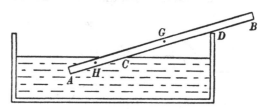

The distances from D of the centre of gravity G and centre of buoyancy H are

$$GD=AD-\tfrac12AB$$

and

$$HD=AD-\tfrac12AC;$$

and the weight of the rod and the force of buoyancy are proportional to σAB and AC. Hence by moments about D

$$\sigma AB(AD-\tfrac12AB)=AC(AD-\tfrac12AC),$$

or

$$\sigma=\frac{AC}{AB}\cdot\frac{2AD-AC}{2AD-AB}\quad\ldots\ldots\ldots\ldots\ldots(1).$$

In the figure the part DB of the rod has a moment about D tending to lift the rod out of the water, so the length immersed will be greatest when AD is greatest, i.e. when B is at D. Then if $AC=xAB$, where $x<1$, (1) becomes

$$\sigma=x(2-x),$$

so that

$$1-\sigma=(1-x)^2,$$

and therefore

$$x=1-\sqrt{(1-\sigma)}.$$

(ii) *A prism of square section floats in water with its long edges horizontal and the centre line of one of its faces hinged to an axis fixed in the surface of the water. Shew that, if the specific gravity of the prism is* $\tfrac{32}{4}$, *the opposite face of the prism will be* $\tfrac34$ *immersed.*   [M. T.]

Let $ABCD$ be the central vertical section of the prism meeting the line of the hinge at O and the water surface at O and E. Let OGF be parallel to AB, G the centre of gravity of the prism, H and H' the centroids of the figures $ODCF$ and OFE. Let a be a side of the square. Then the projection on the horizontal of OG is $\tfrac12a\cos\theta$, of OH is $\tfrac12a\cos\theta-\tfrac14a\sin\theta$, and of OH' is $\tfrac13(OE+OF\cos\theta)$, i.e. $\tfrac16a(\sec\theta+\cos\theta)$. Hence, by equating the moments about O of the weight of the prism and the force of buoyancy, we get

$$\tfrac{32}{4}a^2\cdot\tfrac12a\cos\theta=\tfrac12a^2(\tfrac12a\cos\theta-\tfrac14a\sin\theta)+\tfrac12a^2\tan\theta\cdot\tfrac16a(\sec\theta+\cos\theta),$$

or $\qquad \frac{30}{128} = \frac{1}{2}(2 - \tan \theta) + \frac{1}{6} \tan \theta (\tan^2 \theta + 2);$

i.e. $\qquad \tan^3 \theta + \frac{5}{4} \tan \theta - \frac{21}{64} = 0.$

This equation has one positive root $\tan \theta = \frac{1}{4}$, and for this value $\frac{3}{4}$ of the face BC is immersed.

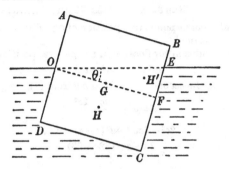

(iii) *A uniform cube of specific gravity s floats in water with two of its faces vertical and one specified edge above the water, the other three hori-zontal edges being immersed; shew that if s lies between $\frac{49}{52}$ and $\frac{1}{2}$, there are three positions of equilibrium.*

Prove that, if $s = \frac{93}{128}$, the cube can float with two of its faces inclined to the vertical at an angle $\tan^{-1} 1·4$.   [M. T.]

Let $ODCE$ be the vertical central section cutting the water surface in AB. Let a denote an edge of the cube. Take OD, OE as axes of x and

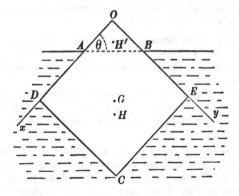

y. Then if G is the centre of the cube, H the centre of buoyancy and H' the centroid of the triangle OAB, the points H', G, H are in a vertical line.

Also if the angle OAB is θ, and $OA = c$, the coordinates of H' are $\frac{1}{3}c$ and $\frac{1}{3}c \tan \theta$, and those of G are $\frac{1}{2}a$, $\frac{1}{2}a$, so that the gradient of $H'G$

is $\dfrac{\frac{1}{2}a - \frac{1}{3}c\tan\theta}{\frac{1}{2}a - \frac{1}{3}c}$, and expressing the fact that this line is at right angles to AB gives
$$\frac{3a - 2c\tan\theta}{3a - 2c} = \cot\theta,$$
which, on reduction, gives
$$\tan\theta = 1 \quad \text{or} \quad (3a - 2c)/2c \quad \dots\dots\dots\dots(1).$$

The first value corresponds to a position of equilibrium in which the diagonal OC is vertical.

The further condition for floating in the general position is
$$sa^2 = a^2 - \tfrac{1}{2}c^2\tan\theta,$$
and, substituting the second value for $\tan\theta$ from (1), we get
$$sa^2 = a^2 - \frac{3ac - 2c^2}{4}$$
or
$$2c^2 - 3ac + 4a^2(1-s) = 0 \quad \dots\dots\dots\dots(2),$$
giving
$$c = \frac{3a \pm a\sqrt{(32s - 23)}}{4}.$$

For real roots in c, we must have
$$s \geqslant \tfrac{23}{32} \quad \dots\dots\dots\dots\dots\dots\dots\dots(3),$$
and for neither root of c/a to exceed unity, we require
$$\sqrt{(32s - 23)} \leqslant 1,$$
or
$$s \leqslant \tfrac{3}{4}.$$

Consequently when s lies between $\tfrac{23}{32}$ and $\tfrac{3}{4}$ there are three possible positions of equilibrium which satisfy the condition that one edge only is above the surface.

Again when $s = \tfrac{93}{128}$, the roots of (2) are $\tfrac{7}{8}a$ and $\tfrac{5}{8}a$, and substituting in (1), we get $\tan\theta = \tfrac{5}{7}$ or $\tfrac{7}{5}$, so that in either case two faces are inclined to the vertical at an angle $\tan^{-1}1\cdot4$.

(iv) *A solid uniform hemisphere of radius a and density σ can turn about O, the centre of its base. The point O is fixed in the surface of a liquid of density ρ. Shew that for a certain value of ρ/σ, the hemisphere will rest in any position.*

If the hemisphere is completely immersed with its centre fixed and is kept at rest with its base vertical by means of a couple, determine, for any values of ρ and σ, the reaction at the centre and the moment of the couple.
[M. T.]

For the solution of this problem we need to know the position of the centre of gravity of a solid lune of a sphere, i.e. a portion cut out by two planes ACB, ADB passing through the same diameter AOB.

The centre of gravity H by symmetry lies in the plane COD at right angles to the diameter AB, and on OE the bisector of the angle COD. Let a be the radius of the sphere and 2α the angle between the planes ACB, ADB. Let the lune be divided up into slices by planes parallel to COD, such as PMR. If the angle ROM is θ, then PMR is a sector of a circle of angle 2α and radius $MR = a\sin\theta$; and the centre of gravity of the sector is on its central radius MQ at a distance $\frac{2}{3}a\sin\theta\dfrac{\sin\alpha}{\alpha}$ from M.*

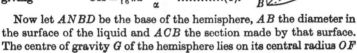

Then if $OM = z$, the thickness of a slice is dz and its area is $\alpha a^2\sin^2\theta$, so that OH is given by

$$OH\int_{-a}^{a}\alpha a^2\sin^2\theta\,dz$$

$$=\int_{-a}^{a}\alpha a^2\sin^2\theta\cdot\tfrac{2}{3}a\sin\theta\frac{\sin\alpha}{\alpha}\,dz.$$

But $z = a\cos\theta$, so that $dz = -a\sin\theta\,d\theta$, and

$$OH\int_{0}^{\pi}\sin^3\theta\,d\theta=\tfrac{2}{3}a\frac{\sin\alpha}{\alpha}\int_{0}^{\pi}\sin^4\theta\,d\theta,$$

giving $OH=\tfrac{3}{16}\pi a\dfrac{\sin\alpha}{\alpha}$ ....(1).

Now let $ANBD$ be the base of the hemisphere, AB the diameter in the surface of the liquid and ACB the section made by that surface. The centre of gravity G of the hemisphere lies on its central radius OF

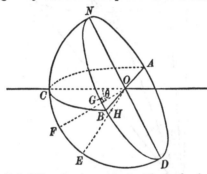

and $OG=\tfrac{3}{8}a$. Let OF make an angle θ with the horizontal. Then the lune immersed is of angle $2\alpha = COD = \tfrac{1}{2}\pi + \theta$, and its volume is $\tfrac{2}{3}a^3\alpha$.

Taking moments about O for the weight and the force of buoyancy, we get

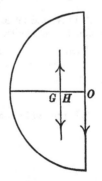

$$\tfrac{2}{3}g\sigma\pi a^3\cdot OG\cos\theta=\tfrac{2}{3}g\rho\alpha a^3\cdot OH\cos\alpha;$$

or from (1)

$$\tfrac{2}{3}g\sigma\pi a^3\cdot\tfrac{3}{8}a\cos\theta=\tfrac{2}{3}g\rho\alpha a^3\cdot\tfrac{3}{16}\pi a\frac{\sin\alpha\cos\alpha}{\alpha},$$

i.e. $\sigma\cos\theta=\rho\sin\alpha\cos\alpha=\tfrac{1}{2}\rho\sin2\alpha=\tfrac{1}{2}\rho\cos\theta.$

It follows that if $\rho = 2\sigma$, equilibrium is possible in any position.

When the hemisphere is completely immersed with its centre O fixed and its base kept vertical, G and H coincide, and the weight $\tfrac{2}{3}g\sigma\pi a^3$ and

* *Statics*, p. 169.

the force of buoyancy $\frac{2}{3}g\rho\pi a^3$, have a resultant $\frac{2}{3}g\,(\rho-\sigma)\,\pi a^3$ acting upwards at G (taking $\rho>\sigma$). The only other force acting on the hemisphere is the reaction at O, so that to maintain equilibrium the reaction at O must be a vertical force $\frac{2}{3}g\,(\rho-\sigma)\,\pi a^3$, and there must also be a couple of moment $\frac{2}{3}g\,(\rho-\sigma)\,\pi a^3 \times \frac{3}{8}a = \frac{1}{4}g\,(\rho-\sigma)\,\pi a^4$.

4·35. Floating cones. In problems on floating right circular cones it is necessary to know how to calculate the volume cut off by a plane oblique to the axis.

Let a plane at a distance c from the vertex O cut the cone in an ellipse of major axis AA' and make an angle $\frac{1}{2}\pi - \theta$ with the axis of the cone. Let 2α be the angle of the cone. Let the perpendicular from O meet

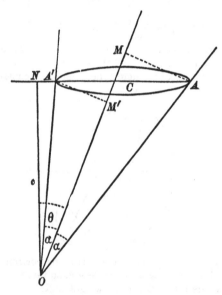

AA' in N, and let M, M' be the projections of A, A' on the axis of the cone. Let a, b be the semiaxes of the ellipse.

Then
$$a = \frac{1}{2}(AN - A'N)$$
$$= \frac{1}{2}c\,\{\tan(\theta+\alpha) - \tan(\theta-\alpha)\}$$
$$= \frac{c\sin 2\alpha}{2\cos(\theta+\alpha)\cos(\theta-\alpha)}.$$

It is proved in books on the geometry of the cone that
$$b = \sqrt{(AM \cdot A'M')}$$
$$= \sqrt{(OA \cdot OA')} \cdot \sin\alpha$$
$$= \frac{c\sin\alpha}{\sqrt{\{\cos(\theta+\alpha)\cos(\theta-\alpha)\}}}.$$

Therefore the area of the ellipse is

$$\pi ab = \pi c^2 \sin^2 \alpha \cos \alpha / \{\cos (\theta + \alpha) \cos (\theta - \alpha)\}^{\frac{3}{2}},$$

and the volume cut off from the cone

$$= \tfrac{1}{3} \pi c^3 \sin^2 \alpha \cos \alpha / \{\cos (\theta + \alpha) \cos (\theta - \alpha)\}^{\frac{3}{2}}.$$

4·351. Example. *Shew that a right circular cone of density ρ and semiangle α can float vertex downwards in a liquid of density σ with one generator vertical and the base just clear of the liquid if*

$$\rho = \sigma (\cos 2\alpha)^{\frac{3}{2}}.$$

Let OAB be the cone, OB a vertical generator, C the centre of the base; and let the surface of the liquid cut the cone in an ellipse of major

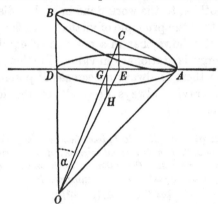

axis AD and centre E. Then the centre of gravity G of the cone is on OC and such that $OG = \tfrac{3}{4} OC$,* and the centre of buoyancy H or centre of gravity of the cone OAD is on OE and such that $OH = \tfrac{3}{4} OE$. Hence HG is parallel to EC; but since E, C are the middle points of DA and BA, therefore EC is parallel to DB which by hypothesis is vertical. Therefore HG is vertical so that one of the conditions for floating in this position is satisfied.

Again from **4·35** the volume OAD

$$= \tfrac{1}{3} \pi OD^3 \sin^2 \alpha \cos \alpha / (\cos 2\alpha)^{\frac{3}{2}};$$

and the volume of the right circular cone OAB

$$= \tfrac{1}{3} \pi OA^3 \sin^2 \alpha \cos \alpha;$$

hence for equilibrium

$$\rho \cdot \tfrac{1}{3} \pi OA^3 \sin^2 \alpha \cos \alpha = \sigma \cdot \tfrac{1}{3} \pi OD^3 \sin^2 \alpha \cos \alpha / (\cos 2\alpha)^{\frac{3}{2}}.$$

But $OD = OA \cos 2\alpha,$

so that the condition becomes

$$\rho = \sigma (\cos 2\alpha)^{\frac{3}{2}}.$$

* *Statics*, p. 165.

4·4. Potential energy of a liquid. A mass of liquid under the action of gravity may be considered to possess potential energy in virtue of its position, like any other material body.* If the liquid be frictionless then in any change of configuration which takes place so slowly that kinetic energy may be neglected, a loss of potential energy will be equal in amount to the work done by gravitational forces and a gain in potential energy will be equal to the work done by some external agent in overcoming gravitational forces. The internal pressures in the liquid belong to the class of 'internal forces' which on the whole contribute nothing to the work done in such a displacement.

Assuming that the principle of virtual work is applicable in such a case, it follows that in a position of equilibrium the potential energy is stationary in value. Further, since in an initial motion there is always a decrease of potential energy, liquid under gravity always descends to the lowest level available to it.

4·41. Example. *A cylindrical piece of wood of length l and sectional area α is floating with its axis vertical in a cylindrical vessel of sectional area A which contains water. Prove that the work done in slowly pressing down the wood until it is just completely immersed is*

$$\tfrac{1}{2}g\alpha l^2 (1-\alpha/A)(\rho-\sigma)^2/\rho,$$

where ρ and σ denote the densities of the water and the wood.   [M. T.]

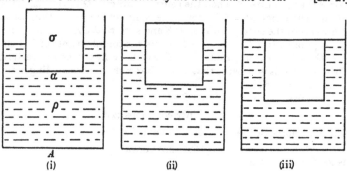

Figs. (i) and (iii) represent the initial and final configurations and fig. (ii) an intermediate configuration.

Initially the length l' immersed is given by

$$\rho l' = \sigma l \quad \dots\dots\dots\dots\dots\dots\dots\dots\dots\dots\dots(1).$$

* *Statics*, p. 211.

In fig. (ii) the block has descended a distance x displacing a volume αx of water which raises the level of the water in the cylinder through a height y given by
$$(A-\alpha)\,y=\alpha x,$$
or
$$\frac{y}{\alpha}=\frac{x}{A-\alpha}=\frac{x+y}{A} \quad\dots\dots\dots\dots\dots\dots(2).$$

In this position the force of buoyancy is
$$g\rho\alpha\,(x+y+l'),$$
and since the weight of the body is $g\sigma\alpha l$, which by $(1)=g\rho\alpha l'$, therefore there is a resultant upthrust equal to
$$g\rho\alpha\,(x+y)$$
or
$$g\rho\alpha Ax/(A-\alpha).$$

To increase the downward displacement by an amount dx, an external agent will therefore have to do work equal to
$$g\rho\alpha Ax\,dx/(A-\alpha).$$

To reach the configuration of fig. (iii), x must increase until
$$x+y+l'=l\dots\dots\dots\dots\dots\dots(3),$$
or, from (1) and (2),
$$\frac{Ax}{A-\alpha}=l\left(1-\frac{\sigma}{\rho}\right),$$
i.e.
$$x=l\left(1-\frac{\alpha}{A}\right)\left(1-\frac{\sigma}{\rho}\right) \quad\dots\dots\dots\dots(4).$$

Denoting this value by x', the total work done
$$=\frac{g\rho\alpha A}{A-\alpha}\int_0^{x'} x\,dx=\frac{1}{2}\frac{g\rho\alpha A}{A-\alpha}x'^2$$
$$=\tfrac{1}{2}g\alpha l^2\left(1-\frac{\alpha}{A}\right)\frac{(\rho-\sigma)^2}{\rho}.$$

Alternatively we may obtain the result by a comparison of the potential energies in configurations (i) and (iii). Thus, if in (iii) the solid has descended a distance x' and the water level has risen a distance y', then x', y' satisfy (2) and (3) and x' is given by (4), and a weight $g\rho\alpha x'$ of water has had its centre of gravity raised through a height
$$l-\tfrac{1}{2}(x'+y'),$$
so that there is a gain of potential energy
$$g\rho\alpha x'\,(l-\tfrac{1}{2}x'-\tfrac{1}{2}y'),$$
which from (1), (3) and (4) is equal to
$$\tfrac{1}{2}g\rho\alpha l^2\left(1-\frac{\alpha}{A}\right)\left(1-\frac{\sigma^2}{\rho^2}\right).$$

But the solid of weight $g\sigma\alpha l$ has descended through a distance x' and therefore has lost potential energy of amount
$$g\sigma\alpha lx'=g\sigma\alpha l^2\left(1-\frac{\alpha}{A}\right)\left(1-\frac{\sigma}{\rho}\right).$$

Consequently the net gain of potential energy or the external work done

$$= \tfrac{1}{2} g \rho \alpha l^2 \left(1 - \frac{\alpha}{A} \right) \left(1 - \frac{\sigma}{\rho} \right)^2,$$

as before.

4·5. Determination of specific gravity. We do not propose to discuss the different methods adopted in laboratories for the determination of specific gravities, but only to call attention to the fact that the Principle of Archimedes provides a method. The specific gravity of a substance being the ratio of the weight of a measured volume of the substance to the weight of an equal volume of a standard substance, say water, the fact that a solid immersed in water is subject to an upward force equal to the weight of the same volume of water suggests the following procedure:

(i) *Neglecting the weight of air displaced.* Let W be the weight of a body A when weighed in air and W' its weight when suspended immersed in water. Then the apparent loss of weight $W - W'$ is a measure of the weight of water displaced, so that the specific gravity is $W/(W - W')$.

If the body does not sink in water, it can be made to sink by attaching to it a sufficiently heavy body B of weight w in air and w' in water; then if the joint body $A + B$ weighs W in air and W' in water, the weight of A alone in air is $W - w$ and in water $W' - w'$, so that the specific gravity of A

$$= (W - w)/(W - w - W' + w').$$

(ii) *Taking into account the weight of the air displaced.* We must assume that the specific gravity of the air is known to be ρ and let σ be the required specific gravity of the body, whose apparent weight in air is W and in water is W'.

If V denotes the volume of the body, we have

$$W = g\sigma V - g\rho V$$

and

$$W' = g\sigma V - g V,$$

so that

$$\frac{W}{W'} = \frac{\sigma - \rho}{\sigma - 1}, \quad \text{and} \quad \sigma = \frac{W - W'\rho}{W - W'}.$$

In general if the weight *in vacuo* of a body of specific gravity

σ is W, then its effective or apparent weight in air of specific gravity ρ is $W\left(1-\dfrac{\rho}{\sigma}\right)$.

Methods for the comparison of the specific gravities of liquids can be devised by the weighing of equal volumes of the liquids, or by weighing in the liquids a solid of known weight in air or vacuo.

4·51. Example. *A graduated glass vessel in the form of a cylinder with vertical walls contains water to a height of 20 cm. from the bottom. A body weighing 20 gm. is placed in it and floats, and the water level rises to 24 cm. The body is then completely submerged, the water level rising to 25 cm. Find the mean specific gravity of the body, the force required to submerge it, and the volume of water in the vessel.*   [M. T.]

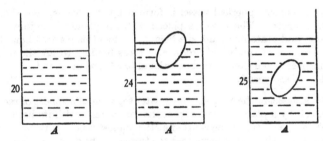

If   the area of the base $= A$ cm.²,

the whole volume of water $= 20A$ c.c.

But $24A$ c.c. $=$ volume of water $+$ immersed volume of solid

and $25A$ c.c. $=$ volume of water $+$ whole volume of solid.

Therefore   the immersed volume of the solid $= 4A$ c.c.

and   the whole volume of the solid $= 5A$ c.c.

Therefore the mean specific gravity of the solid $= \frac{4}{5} = 0.8$.

But the solid weighs 20 gm., therefore when submerged it displaces $20/0.8 = 25$ gm. of water, so that the force necessary to submerge it is the weight of 5 gm.

Again the weight of $4A$ c.c. of water $=$ weight of solid $= 20$ gm.;

therefore $A = 5$ cm.²,

and the volume of water is 100 c.c.

EXAMPLES

1. The bottom of a glass is a circle of an inch diameter, the side forming a portion of a right circular cone of semi-vertical angle 30° with vertex downwards; the glass is filled to a height of 6 inches with water; find approximately in ounces the resultant pressure on the side of the glass, having given that a cubic foot of water weighs 1000 oz. [I.]

2. The stem of a funnel has a radius a and its mouth has a radius b: the height of the stem is h and that of its sloped part is l. The mouth is placed on a horizontal plane, being greased so as to be water-tight: if the funnel is then filled with water to the top of the stem, prove that the upward pressure on the funnel will be equal to

$$\pi w (b-a) \{h (b+a) + \tfrac{1}{3} l (a+2b)\},$$

if w is the weight of unit volume of water. [C.]

3. A hollow spherical vessel is formed by two hemispherical cups joined together. The vessel is placed with the plane of the join horizontal, and contains liquid to a height h above this plane. Shew that the pressure of the liquid produces a force tending to lift the upper hemisphere from the lower proportional to h^3, but independent of the radius of the vessel. [M. T.]

4. A vessel in the shape of the greater segment of a sphere of radius a cut off by a plane distant c from the centre rests with its base on a table, and a vertical tube is fitted at the highest point. If the segment is just filled with water, shew that the thrust on the curved surface will be an upward force if $c < a/2$; and that, if $c > a/2$, the thrust can be made an upward one by pouring water into the tube. [M. T.]

5. A hemisphere of radius a is immersed in a liquid of density σ. The plane of the base is vertical and its centre at a depth $a\sqrt{5}$ below the surface. Shew that the resultant force on the curved surface is $7\pi\sigma g a^3/3$, and that its direction makes an angle θ with the horizontal, where $\tan \theta = 2/\sqrt{45}$. [M. T.]

6. A solid hemisphere (of radius a) is held in liquid with its centre at a depth h. Find the magnitude, direction and position of the resultant thrust (i) on the plane face, (ii) on the curved surface, when the plane face makes an angle θ with the vertical. Find also for what values of θ the magnitude of the latter resultant is greatest and least. [I.]

7. A spherical shell formed of two halves in contact along a vertical plane is filled with water; shew that the resultant pressure on either half of the shell is $\tfrac{1}{4}\sqrt{13}$ of the total weight of the liquid. [I.]

8. A hollow sphere is just filled with heavy homogeneous liquid, so that the pressure at the highest point vanishes. Find the resultant pressure on the lower hemisphere cut off by any central plane; and

shew that as the plane varies, the locus in space of the intersection of the line of action of the pressure with that of the weight of the liquid in this hemisphere is a hemisphere of radius $\frac{2}{3}$ that of the given sphere.

[M. T.]

9. A solid octant of a sphere is immersed in water with a plane face in the surface; prove that the resultant pressure on the curved surface is $\left(1+\frac{8}{\pi^2}\right)^{\frac{1}{2}}$ times the weight of water displaced by the octant. [I.]

10. An octant of a sphere is immersed with one plane side in the surface of the liquid; shew that the depth of the centre of pressure for the curved surface is $\dfrac{\pi a}{(\pi^2+8)^{\frac{1}{2}}}$, where a is the radius of the sphere. [I.]

11. A right cylinder with closed circular ends is filled with water and held in any position. Shew that the resultant thrust on the curved surface bisects the axis at right angles. [M. T.]

12. A solid in the form of an oblique circular cylinder is totally immersed in a homogeneous liquid with its ends horizontal. Shew that the resultant pressure on its curved surface is a couple of moment $Wd \cot \alpha$, where W is the weight and d the depth of the centre of gravity of the liquid displaced and α the inclination of the generators to the planes of the ends. [I.]

13. A right cylinder of any form is completely immersed with its parallel plane ends inclined at an angle α to the horizontal. Shew that the resultant thrust on the cylindrical surface is $\sin \alpha \times$ the weight of the displaced liquid, and determine the line along which it acts. [I.]

14. A surface in the form of a quarter of a circular cylinder of radius a and axial length b is immersed in a fluid of density ρ with its axis and one edge in the free surface. Neglecting atmospheric pressure, shew that the resultant pressure of the fluid on the curved surface is $0\cdot93g\rho a^2 b$ and is inclined at $32\cdot5°$ to the vertical. [M. T.]

15. A right circular cone (height h cm., radius of base a cm.), immersed in water so that its axis is in the free surface, is bisected by a vertical plane. Taking V for the vertex, C for the centre of the base, ACA' and BCB' for the horizontal and vertical diameters of the base, prove that the pressures on the curved surface of one half of the cone, bisected as above stated, are equivalent to three forces, as follows:

(1) A force equal to the weight of $\frac{1}{12}\pi a^2 h$ c.c. of water, acting vertically upwards in a line $\frac{1}{4}h$ cm. from AC and a/π cm. from VC.

(2) A force equal to the weight of $\frac{1}{6}a^2 h$ c.c. of water, acting parallel to AC in a line distant $\frac{3}{8}a$ cm. from VC and $\frac{1}{4}h$ cm. from BC.

(3) A force equal to the weight of $\frac{1}{6}a^3$ c.c. of water, acting parallel to VC in a line distant $\frac{3}{8}a$ cm. from BC and $\frac{3}{16}\pi a$ cm. from AC.

[M. T.]

16. A solid right circular cone of vertical angle 2α is just immersed in water so that one generator is in the surface of the liquid; prove that the resultant pressure on the curved surface of the cone is to the weight of the fluid displaced by the cone as $\sqrt{1+3\sin^2\alpha} : 1$, and that it is inclined to the axis of the cone at an angle $\cot^{-1}(2\tan\alpha)$. [I.]

17. A cone, whose vertical angle is $2a$, has its lowest generator horizontal and is filled with liquid; prove that the resultant pressure on the curved surface is $\sqrt{(1+15\sin^2 a)}$ times the weight of the liquid.

18. A vessel in the form of an oblique cylinder with its base horizontal contains two liquids of densities ρ, σ $(\sigma>\rho)$ of equal depth h. Prove that the thrusts of the liquids on the curved surface are equivalent to a couple of moment

$$\tfrac{1}{2}Wh\cot\alpha\,(3\rho+\sigma)/(\rho+\sigma),$$

where W is the total weight of liquid, and α the inclination of the axis of the cylinder to the base. [M. T.]

19. A rectangular block whose edges are of lengths $2a$, $2b$, $2c$ is divided by a plane through the centre perpendicular to the edges of length $2c$, and the two halves are hinged together along edges parallel to those of length $2a$. The whole is then immersed in a liquid with the line of hinges inclined at an angle θ to the horizon and the dividing plane vertical, the hinges being in the upper face. Prove that the two halves will not separate unless

$$\{(1-\sigma/\rho)\,c^2 - \tfrac{2}{3}b^2\}\cos\theta > 2bd,$$

where d is the depth of the centre of gravity of the block, σ the density of the block and ρ that of the liquid. [M. T.]

20. Two closely fitting hemispheres made of sheet metal of small uniform thickness are hinged together at a point on their rims, and are suspended from the hinge, the rims being greased so that they form a water-tight spherical shell; this shell is now filled with water through a small aperture near the hinge; prove that the contact will not give way if the weight of the shell exceeds three times the weight of water it contains. [M. T.]

21. A surface bounded by a vertical line AB, two equal but not parallel straight lines BC, DA of length a, and a line CD drawn on the surface of a cylinder whose axis is AB and radius a, is in contact with a liquid of density σ; prove that the moment about AB of the resultant thrust is $\tfrac{1}{3}g\sigma a^2(k^2\sim h^2)$, where h, k are the depths of BC and DA. [I.]

22. A hemisphere, totally immersed in a homogeneous fluid, rests with its plane base in contact with a rough fixed inclined plane. Shew that the reactions between the plane and the hemisphere are equivalent to the following forces, viz. a force $\tfrac{1}{4}wh\sigma$ normal to the base and a vertical force $\frac{1}{4}wa\sigma$, both acting at the centre, and a third force equal to the weight of the body acting at its centre of gravity; where σ is the area of the curved surface of the hemisphere, h the depth of its centre, a its radius and w the weight of a unit volume of the fluid. [M. T.]

23. A given mass weighed in air on a spring balance indicates a weight W, it is then compressed to $1/n$th of its former volume and appears to weigh W'; find its weight in vacuo. [M. T.]

24. A ship sailing from a fresh-water river into the sea rises 8 inches, and on then taking extra cargo on board she sinks 6 inches. The original displacement was 1000 tons; shew that the extra cargo weighs 18 tons, taking the specific gravity of salt water to be 1·024. [I.]

25. An iron spherical shell is found to lose half its weight when weighed in water. If the external diameter is 24 inches, determine the thickness of the metal. [Specific gravity of iron = 7·2.]

26. A nugget of quartz and gold weighs 13 oz. in air and 9 oz. in water; taking the specific gravities of quartz and gold to be 2·6 and 19·5 respectively, find the weight of the gold in the nugget.

27. A piece of wax of weight 26·65 grams is tied to a piece of copper; the weight in water of the combination is 8·2 grams. If the weight of the copper in water is 10·25 grams, determine the specific gravity of the wax.

28. A uniform rod has a weight attached to one end to make it float upright in liquid. If 3 inches of the rod is immersed, when it floats in water, and 3·5 inches when it floats in a liquid of specific gravity 0·9, what length of it will be immersed when it floats in a liquid of specific gravity 1·2?

29. A uniform hemispherical shell containing liquid can turn freely about a horizontal axis which passes through two fixed points in its rim, and when it is in equilibrium, the liquid being about to run over, the plane of the rim makes an angle of 45° with the vertical. Shew that the ratio of the weight of the shell to the weight of liquid which would fill the shell is $2c(1-5/4\sqrt{2})/(a-2c)$, where a is the radius of the shell and c the distance of the axis from the centre. [M. T.]

30. The end A of a thin uniform rod AB of length l, mass m and density ρ is attached by a smooth hinge to a point on the base of a vessel containing liquid of density σ, and $\sigma > \rho$. The surface of the liquid is at a height z above A. Shew that, when $z^2 > l^2\rho/\sigma$, the rod can rest only in a vertical position. Shew that, when $0 < z^2 < l^2\rho/\sigma$, the reaction of the hinge is
$$mg(\sqrt{\sigma}-\sqrt{\rho})/\sqrt{\rho}.$$ [M. T.]

31. A cubical box, open at the top and with edges of length a, contains water to a depth b. Find the magnitude and line of action of the resultant of the pressures on a side due to the atmosphere and the water.

A body of weight W is then placed in the water and floats without causing any water to overflow. Prove that the pressure on a side is increased by

$$W\left(\frac{b}{a}+\frac{W}{2W'}\right),$$

where W' is the weight of water that would fill the cube. [I.]

32. A semicircular lamina has one of the ends of its diameter smoothly hinged to a fixed point above the surface of a liquid, and floats with its plane vertical and its diameter half immersed. If the inclination of the diameter to the horizon is $\frac{1}{4}\pi$, prove that the ratio of the density of the liquid to that of the lamina is

$$4\,(3\pi-4):9\pi-8. \qquad \text{[M. T.]}$$

33. A uniform log whose cross-section is a square floats horizontally with one edge in the surface and one edge above the surface of homogeneous liquid. Shew that the ratio of their specific gravities is as 3 is to 4. [M. T.]

34. Two smooth *heterogeneous* hemispheres of equal weights placed base to base form a sphere whose centre of gravity is at its centre; shew that this sphere will remain at rest in any position wholly immersed in homogeneous liquid, whose density is equal to the mean density of the sphere, without the two hemispheres separating. [M. T.]

35. Two liquids, of densities ρ and σ, which do not mix, are contained in a vessel consisting of two vertical cylinders of cross-sections A and B, connected at their lowest parts by a horizontal tube. The free surfaces are respectively at heights a and b above the bottom, and $a>b$. A piece of wood of volume v and density θ (intermediate between ρ and σ) is dropped (1) into the A cylinder, (2) into the B cylinder. Find the changes in the levels of the free surfaces in each case. [M. T.]

36. A sphere of radius a and mass M is loaded so that its centre of gravity G is at a distance c from its centre O, and is suspended by a string attached to a point P of its surface, GP subtending an angle θ at O. The sphere is partly immersed in liquid of density ρ and the tension in the string is $M'g$. Shew that the depth h of O below the surface of the liquid is given by

$$M-M'=\tfrac{1}{3}\pi\rho\,(a+h)^2\,(2a-h),$$

and that the inclination of GO to the vertical is

$$\tan^{-1}\frac{M'a\sin\theta}{Mc-M'a\cos\theta}. \qquad \text{[M. T.]}$$

37. To one end of a thin rod floating horizontally in water is attached a string by which the rod is slowly raised from the water. Shew that, until the vertical position is reached, the tension of the string remains

constant; and that the work done is then $\dfrac{wl(1-s)}{1+\sqrt{(1-s)}}$, where l, w, s are the length, weight and specific gravity of the rod. [M. T.]

38. A rod of length a, weight w, and specific gravity $s\,(<1)$ has a particle of weight W and specific gravity $s'\,(>1)$ fastened at one end; find the value of W so that the rod can rest at any inclination to the vertical, with a length $a\sqrt{s}$ under water. [C.]

39. A cylindrical vessel (A), the area of whose cross-section is α cm.^{2}, is placed with its base on a horizontal table. An iron cylinder (B) whose height is H cm., and specific gravity 7·5, and the area of whose cross-section is β cm.^{2}, rests with its axis vertical on the bottom of A. Mercury (specific gravity 13·5) is now poured into A to a depth h cm. Shew that B will not rise so long as $5H > 9h$. Water is now poured into A until B is immersed. Shew that B will have risen a height

$$(1-\beta/\alpha)(h-13H/25)\,\text{cm.},$$

provided that this expression is positive. [M. T.]

40. If a uniform prism of triangular section floats freely with one edge in the surface of the water, prove that the opposite face must be vertical. [I.]

41. A uniform prism, whose cross-section is an isosceles triangle of vertical angle 2α, floats freely in a liquid with its base just immersed, one edge being in the surface; shew that the ratio of its density to that of the liquid is $2\sin^2\alpha$. [M. T.]

42. A prism, of which the central cross-section is the triangle ABC, has its edge through A horizontal, and is freely movable about this edge which is above the surface of a liquid. The edge through C is below the surface, and AB is horizontal. If the volume immersed be λ^2 of the whole volume, shew that the ratio of the density of the prism to that of the liquid is

$$\lambda^2\{3b\cos A+\lambda(c-2b\cos A)\}:c+b\cos A. \qquad [\text{I.}]$$

43. The cross-section of a uniform prism is an equilateral triangle, and the prism can turn freely about a horizontal axis through the centroids of its cross-sections, which is fixed in the surface of the water. Prove that the couple required to maintain equilibrium in any position is of moment

$$\frac{4aW}{9\sqrt{3}}\frac{\sin\theta\cos^2\theta\,(1-4\sin^2\theta)}{(3-4\sin^2\theta)^2},$$

where a is the length of a side of the cross-section, W the weight of a volume of water equal to that of the prism, and θ the angle which the plane through the axis and the single edge of the prism above or below the surface makes with the vertical. [M. T.]

44. A solid cube made of uniform material can turn freely about one edge which is fixed in the surface of water; prove that if the cube rests

with the face which is not immersed inclined at an angle of 30° to the horizon, the density of the cube is to that of water in the ratio

$$25 - 7\sqrt{3} : 18.$$ [M. T.]

45. A thin-walled hollow cone. of weight W, floats with its axis vertical (vertex upwards) in a fluid of density ρ, being sustained by the buoyancy of the contained air. The angle at the vertex is 90°, and the water level (outside the cone) is midway between the vertex and the surface of the water inside. Neglecting the weight of the contained air, shew that the vertex is at a height h above the outside water level, where $h^3 = 3W/7\pi\rho g$. [M. T.]

46. A cone of semi-vertical angle α and density ρ has its vertex fixed above the surface of a liquid of density σ; if its axis be inclined to the vertical at an angle θ and it floats with one point of its base in the surface of the liquid, prove that

$$\rho = \sigma \left\{ 1 - \frac{\cos\theta \cos^{\frac{2}{3}}(\theta+\alpha)}{\cos\alpha \cos^{\frac{2}{3}}(\theta-\alpha)} \right\}.$$ [I.]

47. A homogeneous tetrahedron floats partially immersed in a liquid so that two of its opposite edges are horizontal. Shew that the straight line joining the middle points of these edges is their shortest distance. [M. T.]

48. A tank is partly filled with water, a space V above the water being empty. A cubical box with thin sides, the weight of which is k times the weight of the water which it can contain, is placed symmetrically in the tank and is gradually depressed by the addition of water inside. Prove that water will begin to overflow from the tank when the depth h of water in the box is given by the equation

$$ka^3 + a^2h = V,$$

provided that $V < a^3$, where a is an edge of the box. What happens if $V > a^3$? [I.]

49. The density of a rod of uniform cross-section varies as the distance from its upper extremity about which, as a fixed point, the rod is free to move. If the rod rests with half of its length in a uniform liquid, shew that the density of the liquid is to the mean density of the rod as 16 is to 9. [C.]

50. A rectangular lamina floats with its plane vertical and its longer side ($2a$) inclined at a *small* angle θ to the vertical. The lamina is entirely immersed in two fluids of densities σ, σ' ($\sigma' > \sigma$), its centre is in the common surface, and its centre of gravity is in the line through the centre parallel to the shortest side ($2b$) and at a small distance c from the centre; prove that

$$\theta(\sigma' - \sigma)(3a^2 - 2b^2) = 6(\sigma' + \sigma)ac.$$ [I.]

51. A sphere of radius a and specific gravity s stops a circular hole of radius b in the vertical wall of a tank containing water, being pressed into the aperture by the pressure of the water but otherwise free to move. Shew that the ball will be thus kept in position by the water if $4a^3s$ lies between the two quantities

$$2a^2(a+c)+b^2c \pm 3b^2z/c,$$

where z is the height of the surface of the water above the centre of the ball (assumed greater than a) and $c^2 = a^2 - b^2$. [M. T.]

52. A right cylinder of height h and specific gravity s is floating in a lake with its axis vertical. Shew that the amount of work which must be done in order to lift it vertically just out of the water is half that required to lift it through the same height in a vacuum. What is the modification in this result when the cylinder floats in a tank of area nA, the area of the section of the cylinder being A? [C.]

53. A sphere, of radius a and specific gravity $\frac{1}{2}$, is held completely immersed at the bottom of a circular cylinder of radius b, which is filled with water to depth d. The sphere is set free and takes up its position of equilibrium; shew that the potential energy lost is

$$W\left(d - \frac{11}{8}a - \frac{1}{3}\frac{a^3}{b^2}\right),$$

where W is the weight of the sphere. [M. T.]

54. A block of stone lies in a tank, with its top in the surface of the water; prove that the work done in lifting the stone slowly just clear of the water is

$$(W - \tfrac{1}{2}W')\,l\,(1 - A/B),$$

where W is the weight of the block, W' of the fluid displaced, A the area of the section of the block, B that of the tank and l is the depth of the block. [C.]

55. A spherical shot of weight W lb. and radius a feet lies at the bottom of a cylindrical bucket, of radius b feet, which is filled up to a depth h feet $(h > 2a)$ with water. Prove that the work done in lifting the shot just clear of the water must exceed

$$W\left(h - \frac{4a^3}{3b^2}\right) - W'\left(h - a - \frac{2a^3}{3b^2}\right) \text{ foot-pounds},$$

the weight of water displaced by the shot being W' lb. [M. T.]

56. A right circular cone is slowly lowered, with its axis vertical, into a cylindrical vessel containing liquid of density ρ, until the base of the cone is just immersed. Shew that the work done on the liquid is $\frac{1}{36}g\rho h^2 B(3A - 2B)/A$, where h is the height and B is the area of the base of the cone, and A the area of the cross-section of the cylinder. [M. T.]

57. A block of wood of density σ, in the form of a right prism of cross-section A and height b, is held with its base in contact with the base of a tank containing liquid of density $\rho\,(>\sigma)$. The liquid in the tank stands at a height h, and covers the block. The block is allowed to rise slowly, performing work, until it floats. If A be so small compared with the horizontal cross-section of the tank that the change of level can be neglected, shew that the work done by the block is

$$gAb\,(\rho-\sigma)\left[h-\tfrac{1}{2}b\left(1+\frac{\sigma}{\rho}\right)\right].$$

If the change of level of the liquid were taken into account, would the work done be found to be less or greater than that given by the above formula? [M. T.]

58. A cylinder of radius $2b$ is filled to depth $8b$ with liquid of density ρ. A solid cylinder of density $\rho/2$, radius b and height $2b$, is then placed in the cylinder and tied to the bottom of the larger cylinder, so that its base is at a height $2b$ above the bottom. Shew that the potential energy of the system when the smaller cylinder is in this position differs from that when it is floating freely by $39g\rho\pi b^4/8$. [M. T.]

ANSWERS

1. 62·5 oz. **6.** (i) $g\rho\pi a^2 h$; (ii) $g\rho\pi a^2\sqrt{\{h^2 \pm \tfrac{4}{3}ah\sin\theta + \tfrac{4}{9}a^2\}}$. Greatest when the base is horizontal and uppermost. Least when the base is horizontal and the curved surface uppermost.

8. $\tfrac{1}{8}g\rho\pi a^3\sqrt{(13+12\sin\theta)}$. **23.** $(nW'-W)/(n-1)$.

25. 1·23 inches. **26.** 3 oz. **27.** $1\tfrac{3}{4}$. **28.** 2·25 inches.

35. The surfaces rise by equal amounts

(1) $v/(A+B)$; (2) $\theta v/\sigma\,(A+B)$.

38. $w\,(1-\sqrt{s})\,s'/(s'-1)\sqrt{s}$. **48.** The box sinks when $h>a\,(1-k)$.

52. The work done $=(n-1)\,wAh^2s^2/2n$, where $w=$ wt. of unit vol. of water. **57.** Less.

Chapter V

STABILITY OF FLOATING BODIES

5·1. Let a body float partially immersed in a liquid. Let G be the centre of gravity of the body and let H be the centre of buoyancy (4·1) in the position of equilibrium. The line GH is then vertical.

When the body undergoes a slight displacement the mass of liquid displaced may be altered in shape or in quantity or in both. An alteration in shape implies a new position H' for the centre of buoyancy, and an alteration in quantity implies a new measure for the force of buoyancy.

Firstly, suppose the body to undergo a small vertical displacement downwards from the equilibrium position, without rotation. This will increase the force of buoyancy so that there will be a resultant upward force on the body tending to restore it to its former position. Similarly an upward vertical displacement would decrease the force of buoyancy and leave a resultant downward force on the body tending to restore it to its former position. Hence the equilibrium of a floating body is *stable for small vertical displacements*.

Secondly, in a small horizontal displacement without rotation, since there is no resultant horizontal force on the body, it would have no tendency to move horizontally from any position in which it is placed. For this type of displacement therefore the equilibrium is neither stable nor unstable.

There only remains for consideration the case of small rotational displacements in which the mass of liquid displaced remains unaltered, and this case we shall consider at length in what follows.

5·2. Metacentre. Let a floating body receive a small rotational displacement which does not alter the mass of liquid displaced. Let G be the centre of gravity of the body and W its weight. Let H be the centre of buoyancy in the equilibrium position and H' the centre of buoyancy in the displaced position. Since the mass of liquid displaced remains unaltered, the force of buoyancy remains equal to W but in the displaced position of the body it acts upwards through H'. If the displacement takes place in a plane of symmetry of the body, the vertical through H' will intersect the line HG which was vertical before the displacement. The point M in which these lines intersect is called the **metacentre** and *the stability of equilibrium depends upon whether the metacentre is above or below the centre of gravity of the body.*

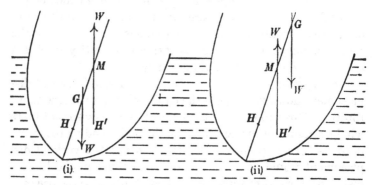

(i)   (ii)

The forces acting on the body are its weight W vertically downwards through G and the force of buoyancy W vertically upwards in the line $H'M$. These forces constitute a couple and in fig. (i), in which M is above G, the couple tends to restore the body to the position in which HG is vertical, which is therefore a stable position in this case; but in fig. (ii) where M is below G the couple tends to increase the inclination of HG to the vertical, so that when M is below G the equilibrium is unstable.

GM is called the **metacentric height** of the floating body. We shall shortly shew how it is to be determined.

If θ is the *small* angular displacement, the couple tending to decrease or increase θ is $W \cdot GM\theta$.

5·21. In general the determination of a metacentre involves integration. If however the immersed surface of the body is spherical, the thrust on every element of the surface passes through the centre of the sphere, and therefore the resultant thrust or force of buoyancy in every position of the body passes through the centre of the sphere. In the same way, if the immersed surface is part of a circular cylinder with horizontal generators, the force of buoyancy will intersect the axis of the cylinder in every position. The following example will serve to shew how we can make use of the foregoing fact in a case in which the immersed surface is not completely spherical.

5·211. Example. *Shew that when a uniform hemisphere of density ρ and radius a floats with its plane base immersed in homogeneous liquid of density σ, the equilibrium is stable and the metacentric height is $\frac{3}{8}a\,(\sigma - \rho)/\rho$.*
[M. T.]

Let CAB be the solid hemisphere and DE its intersection with the surface of the liquid. Imagine the sphere to be completed below the surface of the liquid and let CF be the diameter at right angles to the base AB of the hemisphere.

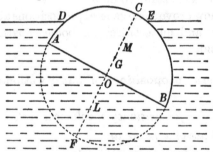

Let V denote the volume immersed, viz. $DABE$. Then the force of buoyancy is an upward force $g\sigma V$ acting through the centre of gravity of $DABE$. Now if we add to the liquid displaced the hemisphere AFB and also subtract it we do not alter the force of buoyancy. But the total upward force will now be the weight of liquid in $DAFBE$, viz. $g\sigma\,(V + \frac{2}{3}\pi a^3)$, acting vertically through O since its components are all normal to the sphere, together with a downward force $\frac{2}{3}g\sigma\pi a^3$ (the added hemisphere) acting vertically through the centre of gravity L of the hemisphere AFB, where $OL = \frac{3}{8}a$.

These two parallel forces have an upward vertical resultant $g\sigma V$ which cuts FC in M, such that

$$g\sigma V \cdot OM = \frac{2}{3}g\sigma\pi a^3 \cdot OL = \frac{1}{4}g\sigma\pi a^4.$$

This point M is the metacentre, for its position is independent of the inclination of FC to the vertical so long as AB is immersed.

And from the condition for floating

$$g\sigma V = \frac{2}{3}g\rho\pi a^3,$$

therefore

$$OM = \frac{3}{8}\frac{\sigma}{\rho}a.$$

But if G is the centre of gravity of the solid hemisphere $OG = \frac{3}{8}a$, and since σ is by hypothesis greater than ρ, therefore M is above G, the position in which GM is vertical is stable, and the metacentric height

$$GM = \tfrac{3}{8}a(\sigma-\rho)/\rho.$$

5·22. In any position of a body floating partially immersed in a liquid the section of the body made by the plane of the surface of the liquid is called **the plane of flotation.**

If a body floating in homogeneous liquid be displaced in such a way that the volume of liquid displaced remains constant, the locus in the body of the centre of buoyancy is called the **surface of buoyancy.**

It will be observed that the surface of buoyancy is the locus of the centre of gravity of a portion of a solid cut off by a plane which cuts off a constant volume.

We shall now prove some general propositions about surfaces of buoyancy and metacentres for bodies floating in homogeneous liquid.

5·3. General propositions about rotational displacements.

   (i) *Condition for constancy of volume and its consequence.*

   (ii) *Existence of a metacentre.*

   (iii) *Formula for metacentric height.*

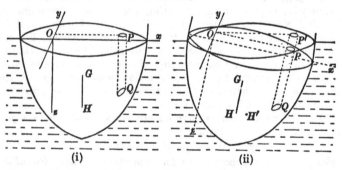

(i)   (ii)

Let $\bar{x}$, $\bar{y}$, $\bar{z}$ be coordinates of the centre of buoyancy H of a body floating in equilibrium with a volume V immersed in homogeneous liquid, when the axes Ox, Oy are in the plane of flotation and Oz vertically downwards (fig. (i)); and let A be the area of the section in the plane of flotation. Suppose the body to be turned through a small angle θ about the axis Oy carrying the axes Ox, Oz with it (fig. (ii)) and let H' be the new centre of buoyancy with coordinates $\bar{x}'$, $\bar{y}'$, $\bar{z}'$. Let G be the centre of gravity of the body.

Then if dA be a small element of area of the surface section at P and z the length of the ordinate PQ between this element and the immersed surface of the body (fig. (i)), the volume immersed is given by

$$V = \int z\, dA \quad \dots\dots\dots\dots\dots\dots(1),$$

where the integration is over the surface section.*

In the displaced position (fig. (ii)) the length of the cylindrical or prismatic element of volume immersed whose cross-section is dA has been increased from PQ to $P'Q$, i.e. from z to $z + x\theta$, to the first order in θ; so that the new volume immersed is represented by

$$\int (z + x\theta)\, dA.$$

If we impose the condition that the volume immersed is to remain constant, we have

$$\int (z + x\theta)\, dA = \int z\, dA,$$

or $$\int x\, dA = 0 \quad \dots\dots\dots\dots\dots\dots(2).$$

This implies that the centroid of the surface section lies on the axis Oy. Hence follows the theorem that

If a plane section of a body cuts off a volume which remains constant for small displacements of the plane, the axis about which the plane turns must pass through the centroid of the section.

Now suppose that the condition of constant volume immersed is satisfied.

---

* The notation implies that the area A is subdivided into a large number of small elements of which dA is a type, that z is the ordinate at some point of dA, that the sum of such products as $z\,dA$ is formed for all the elements of area and then the integral is the limit to which this sum tends when the number of such elements increases and their sizes diminish indefinitely. In actual evaluation it is generally necessary to use double integration in the form $\iint z\,dx\,dy$, but for theoretical purposes the symbolism used and defined here is simpler and unambiguous.

In fig. (i) the coordinates of the centre of gravity of the element PQ or zdA are x, y and $\frac{1}{2}z$, so that the coordinates $\bar{x}$, $\bar{y}$, $\bar{z}$ of H are given by

$$V\bar{x} = \int xz\,dA, \quad V\bar{y} = \int yz\,dA, \quad V\bar{z} = \frac{1}{2}\int z^2\,dA \quad ...(3).$$

And in fig. (ii) the centre of gravity of the element $P'Q$ or $(z + x\theta)\,dA$ is at a distance $\frac{1}{2}(z + x\theta)$ from P' and therefore at a distance $\frac{1}{2}(z - x\theta)$ from P, so that its coordinates are x, y and $\frac{1}{2}(z - x\theta)$ and the coordinates $\bar{x}'$, $\bar{y}'$, $\bar{z}'$ of H' are given by

$$V\bar{x}' = \int x\,(z + x\theta)\,dA, \quad V\bar{y}' = \int y\,(z + x\theta)\,dA, \quad V\bar{z}' = \frac{1}{2}\int(z^2 - x^2\theta^2)\,dA$$
$$......(4).$$

Then by comparing (3) and (4) we see that, to the first power of θ, we have $\bar{z}' = \bar{z}$, so that HH' is parallel to the plane xOy. But H, H' are neighbouring points on the surface of buoyancy and hence follows the theorem:

The tangent plane at any point on the surface of buoyancy is parallel to the corresponding position of the plane of flotation.

We have next to consider under what circumstances a metacentre exists; i.e. what is the condition that the vertical through H' (fig. (ii)) should intersect the line HG which is vertical in equilibrium?

These lines are both parallel to the vertical plane xOz, so that they will intersect if $\bar{y}' = \bar{y}$; i e if

$$\int y\,(z + x\theta)\,dA = \int yz\,dA,$$

or if $\qquad\qquad \int xy\,dA = 0 \qquad ....(5).$

This represents the vanishing of the product of inertia of the area A with regard to the axes Ox, Oy, and this product does vanish if the axis of rotation Oy is a principal axis of the surface section at some point O. This then is the further condition for the existence of a metacentre.

To recapitulate—we have shewn that for small displacements with constant volume immersed the axis of rotation must pass through the centroid of the surface section; there is then a surface of buoyancy and the tangent plane to it at the centre of buoyancy in any such position of the body is horizontal; *and there will be a metacentre for rotation about the axis specified if it is a principal axis of the surface section at some point,* and this is obviously true in any case in which the axis of rotation is an axis of symmetry of the surface section.

It remains, when the foregoing conditions are all satisfied, to find an expression for the metacentric height.

In this case we have

$$HH' = \bar{x}' - \bar{x}$$
$$= \theta \int x^2 dA / V \quad \text{from (3) and (4).}$$

But $\int x^2 dA$ is the moment of inertia of the surface section A about the axis of rotation Oy and may be denoted by Ak^2. Also if M is the metacentre, then HM and $H'M$ are inclined at a small angle θ, so that

$$HH' = HM . \theta.$$

It follows that $\qquad HM = Ak^2 / V \dots\dots\dots\dots\dots(6).$

We observe that this expression for the height of the metacentre above the centre of buoyancy depends only on the surface section and the total volume immersed.

The **metacentric height** GM (5·2) is given by

$$GM = HM - HG = (Ak^2 - V . HG)/V \ \dots\dots\dots(7);$$

and if ρ is the density of the liquid so that the weight W of the body is equal to $g\rho V$, the couple tending to restore equilibrium is

$$W . GM . \theta = g\rho\theta (Ak^2 - V . HG) \ \dots\dots\dots\dots(8).$$

5·31. Since in general there are two principal axes of the surface section of a floating body passing through its centroid, with corresponding moments of inertia $Ak_1{}^2$ and $Ak_2{}^2$, therefore there are two metacentres, one for rotation about each of these axes; and the equilibrium will be stable for displacements about both the axes if $HG < Ak_1{}^2/V$ and also $< Ak_2{}^2/V$.

It can be shewn that the equilibrium is stable for displacements about *all* horizontal axes through the centroid of the surface section if the centre of gravity of the body is lower than the lower of these two metacentres. For a proof of this theorem in the general case we have to make use of the general equation of the surface of buoyancy.*

* Besant and Ramsey, *A Treatise on Hydromechanics*, Part I, *Hydrostatics*, § 66.

5·32. The work done in producing a small displacement of the kind described in **5·2** and **5·3** is obtained from the consideration that, when the angular displacement is θ, the restorative couple is $W . GM . \theta$, so that to increase the displacement by an amount $d\theta$ would require an amount of work $W . GM . \theta d\theta$, and, by integration with regard to θ between limits 0 and θ, the work required to produce the displacement θ is

$$\tfrac{1}{2} \dot{W} . GM . \theta^2$$

or

$$\tfrac{1}{2} g\rho \, (Ak^2 - V . HG) \, \theta^2.$$

5·33. A practical method of determining the metacentric height of a ship consists in measuring the relative deflection of a plumb line produced by moving a known weight w through a measured distance d across the deck. Since the removal of a

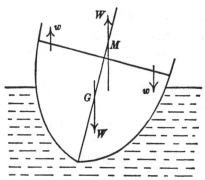

force w from one point and its application at another is equivalent to the application of a couple, therefore if the vessel turns through a small angle θ into a new position of equilibrium, we have, by equating the couples,

$$wd \cos \theta = W . GM . \sin \theta,$$

or, θ being small, $GM = wd/W\theta.$

5·34. Examples. (i) *Discuss the stability of a uniform right circular cone of density σ floating in a liquid of density ρ with its axis vertical and vertex downwards.*

Let h be the height of the cone, 2α its angle and h' the length of axis immersed.

In this case A is a circle of radius $h' \tan \alpha$ so that

$$Ak^2 = \tfrac{1}{4} \pi h'^4 \tan^4 \alpha *$$

* See the 'Reference Table of Moments of Inertia', *Dynamics*, Pt. I, p. 189.

and $V = \frac{1}{3}\pi h'^3 \tan^2\alpha,$

therefore $HM = \frac{3}{4}h' \tan^2\alpha.$

But, if O is the vertex, $OH = \frac{3}{4}h'$, so that $OM = \frac{3}{4}h' \sec^2\alpha.$

But $OG = \frac{3}{4}h.$ Therefore the equilibrium is stable or unstable according as

$$h' \sec^2\alpha > \text{ or } < h.$$

But, since the cone floats, $\rho h'^3 = \sigma h^3$, so that the equilibrium is stable or unstable according as

$$\sigma/\rho > \text{ or } < \cos^6\alpha.$$

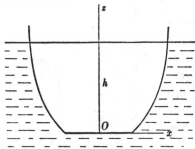

(ii) *Find the form of a uniform solid of revolution such that when floating with its axis vertical the distance between the centre of buoyancy and the metacentre is independent of the length of the axis immersed.*

Let $x^2 = f'(z)$ ....(1)

be the equation of a meridian section of the surface, where the origin O is in the lowest section and Oz is measured upwards, and let h be the length of the axis immersed.

In this case A is a circle of radius $\sqrt{(f'(h))}$, so that

$$Ak^2 = \frac{1}{4}\pi f'(h)^2.$$

And $V = \int_0^h \pi x^2\, dz = \pi \int_0^h f'(z)\, dz = \pi\{f(h) - f(0)\}.$

But $Ak^2 = cV$, where c is a constant.

Therefore $\qquad\qquad f'(h)^2 = 4c\{f(h) - f(0)\}$ ....(2)

or $\qquad\qquad\qquad \dfrac{f'(h)}{2\sqrt{\{f(h) - f(0)\}}} = \sqrt{c}.$

Hence by integration

$$\sqrt{\{f(h) - f(0)\}} = \sqrt{c}\,h + b \qquad ....(3),$$

where b is a constant of integration.

But the left-hand side of (3) vanishes when $h = 0$, so that $b = 0$.

Hence $\qquad\qquad f(h) = f(0) + ch^2$ ....(4)

and $\qquad\qquad\qquad f'(h) = 2ch,$

so that $\qquad\qquad\qquad f'(z) = 2cz;$

and from (1) the required equation of the meridian curve is

$$x^2 = 2cz,$$

and the surface is a paraboloid of revolution.

(iii) *A homogeneous circular cylinder of length h, radius a, and specific gravity ρ floats in water. Prove that the position with the axis vertical is stable if $a^2/h^2 > 2\rho(1 - \rho)$; also that the position with the axis horizontal is stable if $h > b$, where b is the breadth of the rectangular water section.*

Prove that, if $\rho = \frac{1}{2}$, and $8a^2 = 3h^2$, the position of stable equilibrium is one in which one end of the cylinder is just immersed, and the other is just out of the water. [I.]

Let h' be the length immersed when the axis is vertical. Then $h' = \rho h$. Also, if O is the centre of the base, $OG = \frac{1}{2}h$ and $OH = \frac{1}{2}h'$; $Ak^2 = \frac{1}{4}\pi a^4$ and $V = \pi a^2 h'$.

The condition for stability is $OM > OG$, or $OH + HM > OG$, i.e.

$$\tfrac{1}{2}h' + \frac{a^2}{4h'} > \tfrac{1}{2}h,$$

i.e. $\qquad\qquad \rho h + \dfrac{a^2}{2\rho h} > h, \quad$ or $\quad a^2/h^2 > 2\rho(1 - \rho).$

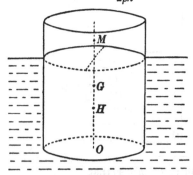

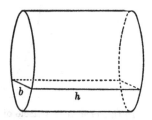

When the axis is horizontal, for rolling displacements the force of buoyancy always intersects the axis, so that the metacentre M_1 lies on the axis. Since the section by the water surface is a rectangle of sides b and h, the moment of inertia Ak^2 for tilting will be greater than for rolling if $h > b$. In this case therefore the metacentre M_2 for tilting will be higher than the metacentre for rolling and the equilibrium will be stable for both types of displacements (5·31), or rather neutral for rolling and stable for tilting.

When $\rho = \frac{1}{2}$, G lies in the plane of flotation. Consider the position in which one end AB is just immersed and the other end CD just out of the water. Take O as origin, the axis Ox along the line of greatest slope in the base, and Oz along the axis of the cone. Then, if the base is

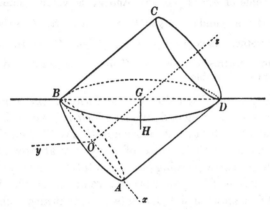

inclined at an angle α to the horizontal, the plane of the water section is the plane $z = (x + a)\tan\alpha$. The centre of buoyancy H lies in the plane xOz and its coordinates $\bar{x}$, $\bar{z}$ are given by

$$\bar{x}\iint z\,dx\,dy = \iint xz\,dx\,dy, \quad \bar{z}\iint z\,dx\,dy = \iint \tfrac{1}{2}z^2\,dx\,dy,$$

where $z\,dx\,dy$ represents a prismatic element of volume standing on a base $dx\,dy$, of height $z = (x + a)\tan\alpha$, and the integration is over the circular base. Hence

$$\bar{x}\iint (x + a)\,dx\,dy = \iint (x^2 + ax)\,dx\,dy$$

and $\bar{z}\iint (x + a)\,dx\,dy = \tfrac{1}{2}\tan\alpha \iint (x^2 + 2ax + a^2)\,dx\,dy.$

But, since the origin is the centroid of the area of integration, therefore $\iint x\,dx\,dy = 0$, while $\iint dx\,dy = \pi a^2$, and $\iint x^2\,dx\,dy = \tfrac{1}{4}\pi a^4$, the last integral being the moment of inertia of a circle about a diameter.

Hence $\bar{x}.\pi a^3 = \tfrac{1}{4}\pi a^4,$ or $\bar{x} = \tfrac{1}{4}a,$

and $\bar{z}.\pi a^3 = \tfrac{5}{8}\pi a^4 \tan\alpha,$ or $\bar{z} = \tfrac{5}{8}a\tan\alpha.$

Therefore $\tan OGH = \tfrac{1}{4}a/(\tfrac{1}{2}h - \tfrac{5}{8}a\tan\alpha),$

or since $h = 2a\tan\alpha,$ $\tan OGH = \tfrac{2}{3}\cot\alpha.$

But for GH to be vertical we must have $OGH = \alpha$, and therefore

$$\tan^2\alpha = \tfrac{2}{3}, \quad \text{or} \quad \frac{h^2}{4a^2} = \tfrac{2}{3}, \quad \text{i.e. } 3h^2 = 8a^2.$$

Also in this position

$$HG = \bar{x} \operatorname{cosec} \alpha = \tfrac{1}{4} a \operatorname{cosec} \alpha = \frac{\sqrt{5}}{4\sqrt{2}} a.$$

Again $HM = Ak^2/V$, where A is the area of the water section, i.e. an ellipse, and k is its radius of gyration about the axis about which displacement takes place. The smallest moment of inertia is about the major axis BD, and since the minor axis is a diameter of the circular section, the least k^2 is $\tfrac{1}{4} a^2$. Also, by projection, $A = \pi a^2 \sec \alpha$, so that the least value of Ak^2 is $\dfrac{\sqrt{5}}{4\sqrt{3}} \pi a^4$. And V, the volume immersed, is half that of the cylinder, i.e. $\tfrac{1}{2} \pi a^2 h$ or $\sqrt{\tfrac{2}{3}} \pi a^3$. Therefore the corresponding metacentre is given by $HM = \dfrac{\sqrt{5}}{4\sqrt{2}} a$. Hence the lower of the two metacentres coincides with G and the higher is above G and the equilibrium is stable.

5·4. A general discussion of the form of the surface of buoyancy for a floating body of given form is beyond the scope of an elementary text book, but in addition to the fact that the tangent plane to the surface of buoyancy at any point is parallel to the corresponding plane of flotation, we may observe that the surface of buoyancy is concave upwards. This follows from the fact shewn in **5·3** (3) and (4), viz. that although to the first power of θ we have $\bar{z}' = \bar{z}$, yet, when we do not neglect θ^2, $\bar{z}'$ is less than $\bar{z}$ by the amount $\tfrac{1}{2}\theta^2 \int x^2 dA/V$ or $\tfrac{1}{2}\theta^2 Ak^2/V$ which is essentially positive, so that all points on the surface of buoyancy in the immediate neighbourhood of H are nearer to the plane of flotation than H.

Again, when the surface of buoyancy is known, for a given body floating in a liquid of given density, possible positions of equilibrium of the body may be found by drawing normals to the surface of buoyancy from the centre of gravity G of the body. For if GH be such a normal and the body be placed in the liquid with GH vertical and the specified volume immersed, the tangent plane at H to the surface of buoyancy will be horizontal and all the necessary conditions for floating are satisfied.

Further, if H' is the centre of buoyancy in a slightly displaced position the vertical through H' is also a normal to the surface of buoyancy, and if the displacement is such that there is a metacentre M, then M is the point of ultimate intersection of two neighbouring normals, i.e. M is the centre of curvature of one of the principal normal sections of the surface of buoyancy; and for sufficiently near points H, H' we have $HM = H'M$.

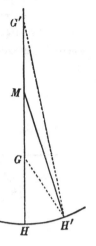

Now suppose that GH is a normal from the centre of gravity to the surface of buoyancy, and that G lies between H and the meta-centre M. Then GH being a normal is stationary in length for small displacements of H. But $H'G + GM > H'M$, and $H'M = HM$, therefore $H'G + GM > HM$, or $H'G > HG$. So that in this case HG is a minimum. But by 5·2 this is the case of stable equilibrium. Hence for stable equilibrium HG is a minimum. In like manner we can shew that when the centre of gravity lies on HM produced, as at G' in the figure, $HG' > H'G'$, or HG' is a maximum, so that in unstable equilibrium HG is a maximum.

5·41. We shall illustrate the method of **5·4** by applying it to a simple case, assuming an elementary knowledge of analytical geometry of three dimensions.

Find the surface of buoyancy for a uniform rectangular solid partially immersed in liquid with one face completely immersed.

A cube floats freely in liquid of twice its density; shew that there are an infinite number of positions of stable equilibrium in which the water line is a parallelogram, and such that the faces entirely in and out of the fluid are inclined at an angle of $\frac{1}{4}\pi$ to the surface. [I.]

Let $ABCD$ be the lowest face of the solid and $PQRS$ the water line.

Take an origin O at the centre of $ABCD$ and axes Ox, Oy in that plane parallel to the edges, and Oz at right angles to the plane. Let the equation of the water surface be

$$z = lx + my + c \quad \dots\dots\dots\dots\dots\dots(1).$$

This plane cuts Oz in N where $ON = c$, and since the volume of the solid cut off by all planes through N is the same, therefore the plane of flotation always passes through the point N.

Let $AB = 2a$, $BC = 2b$ and let V be the volume immersed, and $\bar{x}$, $\bar{y}$, $\bar{z}$ the coordinates of the centre of buoyancy H.

Then integrating by summing prismatic elements of volume of section $dx\,dy$ and height z, where z is given by (1), we have

$$V = \iint (lx + my + c)\,dx\,dy,$$

$$V\bar{x} = \iint x\,(lx + my + c)\,dx\,dy, \quad V\bar{y} = \iint y\,(lx + my + c)\,dx\,dy$$

and $$V\bar{z} = \iint \tfrac{1}{2}z^2 dx\,dy = \tfrac{1}{2}\iint (lx + my + c)^2\,dx\,dy;$$

where the integrations are over the rectangle $ABCD$.

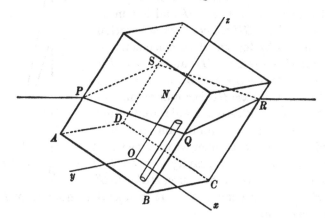

Since the axes are the principal axes of the area $ABCD$ at its centre of gravity, therefore

$$\iint x\,dx\,dy = \iint y\,dx\,dy = \iint xy\,dx\,dy = 0;$$

and $$\iint x^2 dx\,dy = 4ab\cdot\frac{a^2}{3}, \quad \iint y^2 dx\,dy = 4ab\cdot\frac{b^2}{3}.$$

Hence $V = 4abc$ is independent of l and m as stated above,

$$V\bar{x} = 4ab\cdot\frac{a^2}{3}\,l, \quad V\bar{y} = 4ab\cdot\frac{b^2}{3}\,m \quad\ldots\ldots\ldots\ldots\ldots(2)$$

and $$V\bar{z} = 2ab\left(\frac{a^2}{3}\,l^2 + \frac{b^2}{3}\,m^2 + c^2\right) \quad\ldots\ldots\ldots\ldots(3).$$

By eliminating V, l, m, we find that $\bar{x}$, $\bar{y}$, $\bar{z}$ satisfy the equation

$$\frac{x^2}{a^2} + \frac{y^2}{b^2} = \frac{2}{3}\left(\frac{z}{c} - \frac{1}{2}\right) \quad\ldots\ldots\ldots\ldots\ldots(4),$$

and this is the equation of the surface of buoyancy. It is a paraboloid and clearly passes through the point $(0, 0, \tfrac{1}{2}c)$, which is the position of H when the base $ABCD$ is horizontal.

For the second part of the question, if the solid is a cube of side $2a$, immersed in fluid of twice its density, then $a = b = c$, and equations (1) and (4) become

$$z = lx + my + a \quad\ldots\ldots\ldots\ldots\ldots\ldots\ldots(1')$$

and $$x^2 + y^2 = \tfrac{2}{3}a\,(z - \tfrac{1}{2}a) \quad\ldots\ldots\ldots\ldots\ldots\ldots(4').$$

The coordinates of the centre of gravity G are $(0, 0, a)$.

Now the equations of the normal at any point $(\bar{x}, \bar{y}, \bar{z})$ on the surface $f(x, y, z) = 0$, being

$$\frac{x-\bar{x}}{\partial f/\partial \bar{x}} = \frac{y-\bar{y}}{\partial f/\partial \bar{y}} = \frac{z-\bar{z}}{\partial f/\partial \bar{z}},$$

it follows that the normal at a point H $(\bar{x}, \bar{y}, \bar{z})$ on the surface $(4')$ is

$$\frac{x-\bar{x}}{\bar{x}} = \frac{y-\bar{y}}{\bar{y}} = \frac{z-\bar{z}}{-\frac{1}{3}a} \quad \dots\dots\dots\dots\dots(5),$$

and this line passes through the point G, $(0, 0, a)$, if

$$\frac{-\bar{x}}{\bar{x}} = \frac{-\bar{y}}{\bar{y}} = \frac{a-\bar{z}}{-\frac{1}{3}a},$$

i.e. if H lies on the plane $\quad z = \frac{2}{3}a \quad \dots\dots\dots\dots\dots\dots(6).$

Hence normals from G to the surface of buoyancy $(4')$ meet it on the plane $z = \frac{2}{3}a$, which cuts $(4')$ in the circle

$$x^2 + y^2 = \frac{1}{9}a^2 \dots\dots\dots\dots\dots\dots(7).$$

Now the plane $(1')$ being horizontal, the direction cosines of the vertical are proportional to $-l, -m, 1$, so that the cosine of the angle between the vertical and Oz or between the plane $ABCD$ and the horizontal is given by

$$\cos\theta = \frac{1}{\sqrt{(l^2+m^2+1)}} \quad \dots\dots\dots\dots\dots(8).$$

But from (2), in this case $l = 3\bar{x}/a$ and $m = 3\bar{y}/a$, so that

$$l^2 + m^2 = 9(\bar{x}^2 + \bar{y}^2)/a^2 = 1, \quad \text{(from (7))}.$$

Therefore $\quad\quad \cos\theta = \frac{1}{\sqrt{2}} \quad$ or $\quad \theta = \frac{1}{4}\pi.$

Since the plane (6) cuts the surface $(4')$ in a circle, there are therefore infinitely many positions for H satisfying the condition that the normal at H goes through G and the plane $ABCD$ is inclined at $\frac{1}{4}\pi$ to the horizontal.

To consider the stability we have

$$HG^2 = \bar{x}^2 + \bar{y}^2 + (a-\bar{z})^2,$$

and in a displaced position $(\bar{x}, \bar{y}, \bar{z})$ satisfy $(4')$; so that

$$HG^2 = \frac{2}{3}a(\bar{z}-\frac{1}{3}a) + (a-\bar{z})^2.$$

Denoting this by $f(\bar{z})$, we have

$$f'(\bar{z}) = \frac{2}{3}a - 2(a-\bar{z});$$

this vanishes for $\bar{z} = \frac{2}{3}a$, in agreement with (6), and this value makes $f''(\bar{z})$ positive so that HG is a minimum and the equilibrium is stable.

5·42. *A solid of uniform density σ floats partly immersed in a homogeneous liquid of density ρ. Shew that a solid of the same size and shape and of uniform density $\rho - \sigma$ can float inverted with the same plane of flotation in the same liquid; and that if the equili-*

brium is stable in the one case it is also stable in the other case for like displacements.

Let V, V' denote the volumes immersed and unimmersed in the first position of equilibrium and H, H' the centroids of V and V'.

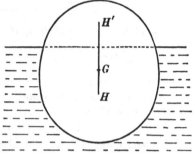

Then if G is the centre of gravity, HH' passes through G; but HG is vertical in the first case, therefore $H'G$ is vertical when the body is inverted. Again $\dfrac{V}{\sigma} = \dfrac{V+V'}{\rho}$, therefore also $= \dfrac{V'}{\rho - \sigma}$, or $V'\rho = (V+V')(\rho - \sigma)$; so that both conditions of equilibrium are satisfied in the second case.

With the usual notation, the first position is stable if $HM > HG$, i.e. if $Ak^2 > V \cdot HG$. But $V \cdot HG = V' \cdot H'G$, so that we also have $Ak^2 > V' \cdot H'G$ and this is the condition for stability in the second case.

When the liquid is water, then if s is the specific gravity of the first solid $1 - s$ is that of the second. It follows that in considering the limits to the specific gravity of a solid of given form in order that it may float in stable equilibrium we may confine ourselves to values of s less than $\frac{1}{2}$, since for any value of s less than $\frac{1}{2}$ which ensures stability there is a corresponding value $1 - s$ which also gives a stable position.

5·5. Vessel containing liquid. When homogeneous liquid is contained in a vessel which undergoes a small rotational displacement the determination of the line of action of the resultant *downward* thrust on the vessel, which is of course the weight of the liquid, is analogous to the determination of the line of action of the upward thrust on a floating body. For the liquid in the vessel occupies a given volume, and the free surface of the liquid remains horizontal, so that figs. (i) and (ii) of **5·3** might represent the vessel of liquid in two neighbouring positions. The line of action of the resultant thrust in the displaced position is the vertical through H', which, in cases of symmetry, will cut HG in a metacentre M determined as in **5·3**.

5·51. Floating vessel containing liquid. To take a simple case, let the form of the floating vessel be a figure of revolution floating with its axis vertical and containing some liquid. There is then symmetry about all vertical planes through the axis.

Let W, W' be the weights of the displaced and the contained liquid. There is a metacentre M for the line of action of the force of buoyancy W determined as in **5·3**, and there is

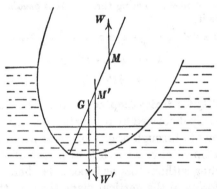

likewise a metacentre M' for the line of action of the weight W'. The only other vertical force is the weight of the vessel acting through its centre of gravity G. The equilibrium will be stable when the axis is vertical if in the displaced position there is a resultant moment about G tending to restore the vertical position, i.e. if

$$W \cdot GM - W' \cdot GM'$$

is positive, or if

$$\frac{W}{W'} > \frac{GM'}{GM}.$$

5·52. Example. *A thin metal circular cylinder contains water to a depth h and floats in water with its axis vertical immersed to a depth h'. Shew that the vertical position is stable if the height of the centre of gravity of the cylinder above its base is less than* $\frac{1}{2}(h+h')$.

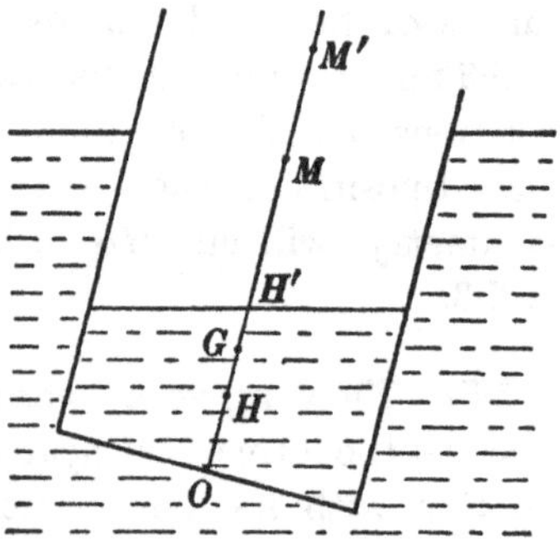

Let G be the centre of gravity of the cylinder, O the centre of the base, and H, H' the centres of gravity of the water contained and displaced in the equilibrium position of the cylinder; M, M' the corresponding metacentres, a the radius of the cylinder and let $OG = z$.

Then

$$H'M' = Ak^2/V$$
$$= \tfrac{1}{4}\pi a^4/\pi a^2 h' = a^2/4h',$$

so that

$$GM' = OH' + H'M' - OG = \frac{h'}{2} + \frac{a^2}{4h'} - z.$$

Similarly

$$GM = \frac{h}{2} + \frac{a^2}{4h} - z.$$

The upward force of buoyancy acting through M' is $g\rho\pi a^2 h'$ and the weight of water contained acting through M is $g\rho\pi a^2 h$, so the equilibrium is stable if

$$g\rho\pi a^2 h' \left(\frac{h'}{2} + \frac{a^2}{4h'} - z\right) > g\rho\pi a^2 h \left(\frac{h}{2} + \frac{a^2}{4h} - z\right),$$

or if

$$\tfrac{1}{2}(h'^2 - h^2) > z(h' - h),$$

or

$$z < \tfrac{1}{2}(h + h').$$

5·6. Stability for finite displacements. In considering the stability of a ship displaced by rolling it is necessary to provide for displacements through angles of finite magnitude. If we assume the ship to have two vertical planes of symmetry so that pure rolling without tossing is possible, then we have only to consider forces in the vertical plane through the centre of gravity G transverse to the length of the ship. The section of the surface of buoyancy by this plane is *the curve of buoyancy*—the curve HP in the figures, where H is the centre of buoyancy in the position of equilibrium. This curve is symmetrical about the line HG, so that its evolute (locus of centres of curvature or envelope of normals) has a cusp at M_0 on the line HG, M_0 being the metacentre in the equilibrium position; and the cusp may point downwards as in fig. (i) or upwards as in fig. (ii). If the vessel turns through an angle θ so that the normal PM to the curve of buoyancy becomes vertical, then the force of buoyancy acts along PM and the restorative couple is $W . GY$, where W is the weight of liquid displaced and GY is perpendicular to PM. In fig. (i) it is evident that GY increases steadily with θ, so that the equilibrium is stable when the evolute of the curve of buoyancy is as in fig. (i). But in fig. (ii),

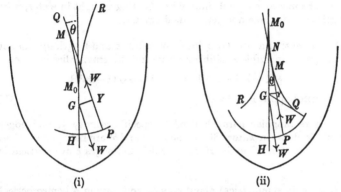

(i)   (ii)

if the normal PM to the curve of buoyancy cuts HG in N, it is clear that as θ increases GN decreases until the normal passes through G, when GN vanishes, so that GY increases to a maximum and then decreases to zero, and the position in which the normal passes through G would be a position of equilibrium, but an unstable one because as θ increases beyond the value corresponding to GQ the couple ceases to be restorative and tends to increase the displacement. Hence with the evolute of the curve of buoyancy as in fig. (ii) the equilibrium becomes unstable for sufficiently large displacements.

EXAMPLES

1. A uniform circular cylinder, whose radius is two-thirds of its height, floats in water with its axis vertical. Prove that the equilibrium cannot be stable if the specific gravity of the cylinder lies between $\frac{1}{3}$ and $\frac{2}{3}$.

2. A thin cylindrical stick consists of two parts equal in length. The specific gravity of one part is $\frac{5}{8}$. Shew that the least specific gravity of the other part that will permit of the stick floating upright in stable equilibrium in water is $\frac{1}{9} \cdot 0$.

3. A thin rectangular strip of uniform material, having sides of length l and πa, is bent into a channel of semi-circular section (radius a and length l) with open ends and immersed in water. The specific gravity of the material is $\frac{1}{2}$. Confining attention to positions in which the straight edges of the channel are horizontal, shew that there are two configurations of stable equilibrium. Are there any configurations of unstable equilibrium? [M. T.]

4. Shew that a uniform circular cylinder, of specific gravity $\frac{1}{2}$, cannot be in stable equilibrium, when floating upright in water, if its length exceeds three-quarters of its diameter.

5. Prove that a circular cylinder of radius a and length a/n cannot float upright in stable equilibrium if its specific gravity lies between

$$\tfrac{1}{2}[1-\sqrt{(1-2n^2)}] \quad \text{and} \quad \tfrac{1}{2}[1+\sqrt{(1-2n^2)}].$$

What will happen if $2n^2 > 1$? [C.]

6. A rectangular parallelepiped of specific gravity $\frac{1}{2}$, and edges $2a$, $2b$, $2c$, floats in water. Shew that, if a^2 and b^2 are both greater than $\frac{3}{2}c^2$, the position of equilibrium in which the edges c are vertical is stable.

7. A body symmetrical about an axis and having a hemispherical base rests on a rough horizontal table, and the equilibrium is unstable. Shew that if the body has in it a spherical cavity, the equilibrium may be rendered stable by introducing liquid, provided the centre of the cavity is below the centre of the hemisphere, but otherwise not. Give a formula for the smallest mass of liquid that will suffice. [C.]

8. A cylindrical cup is formed of thin sheet metal, the height being twice the diameter; the surface density of the plane bottom is n times that of the curved surface, and the weight of the cup is half that of the water which would fill it. Shew that the cup will float in stable equilibrium with its generators vertical, if $n > \frac{5}{9}$. [M. T.]

9. If a segment of a sphere of density σ floats in liquid of density ρ, prove that the position in which the plane face is immersed and horizontal is a position of stable equilibrium and that the metacentric height is $\dfrac{\rho - \sigma}{\sigma}$ times the distance of the centre of gravity of the segment from the centre of the sphere. [I.]

10. Shew that a homogeneous right circular cone of vertical angle 2α cannot float stably with its axis vertical and vertex downwards unless its density as compared with that of the liquid is greater than $\cos^6 \alpha$.

What is the corresponding result when the vertex is upwards? [C.]

11. If in the first part of the preceding example $\cos^6\alpha > \sigma/\rho$, prove that the position becomes stable when a particle of weight w is attached to the vertex if

$$\frac{w}{W} > \left\{\left(\frac{\rho\cos^6\alpha}{\sigma}\right)^{\frac{1}{4}} - 1\right\},$$

where W is the weight of the cone. It is assumed that the base of the cone is not submerged, that σ is the density of the cone and ρ that of the liquid. [I.]

12. Shew that for a solid cone (not necessarily circular) floating vertex downwards the surfaces of buoyancy and of flotation are similar and similarly situated, and that their linear dimensions are in the ratio 3 : 4. [C.]

13. A uniform elliptic disc of eccentricity e and mass M has a particle of mass m attached at one focus and floats with its major axis vertical in a fluid of such density that one-fourth of that axis is immersed. Shew that the equilibrium will be stable if

$$(4\pi/3\sqrt{3} - 1 - e)m > eM. \qquad [C.]$$

14. A homogeneous cylinder whose cross-section is an ellipse floats with its generators horizontal. Shew that one of the principal axes must be vertical. Which one is vertical in stable equilibrium? [C.]

15. Investigate the stability of the different positions in which an ellipsoid can float in a liquid of twice its specific gravity. [I.]

16. A hollow circular cylinder floats in water with its axis vertical, and three quarters immersed, prove that for stability the ratio of the radius to the height of the cylinder must be greater than $\sqrt{\frac{3}{8}}$. Investigate the effect on the stability of putting a given quantity of water into the cylinder. [I.]

17. A thin cylindrical vessel floats when empty with its axis vertical and its base a depth h below the surface, the equilibrium being unstable. Shew that the vessel can float in stable equilibrium with its axis vertical if water of volume greater than $\pi a^2(c - h/2)$ is poured in; a is the radius of the vessel and c the distance of its centre of gravity from its base. [I.]

18. A flat-bottomed, vertical-sided ship of waterline area A and weight W floats normally with its bottom horizontal. Enough water leaks in to keep the bottom covered. Neglecting the thickness of the sides, shew that the ship will not remain vertical if $h > \dfrac{W}{2g\rho A}$, where ρ is the density of water, and h is the height of the centre of gravity of the ship above the water inside. [I.]

19. A thin circular hollow cylinder floats in water with its generators vertical, and also contains water. Shew that the condition of stability is that the centre of gravity of the cylinder should be below the plane midway between the internal and external water lines.

20. A vase is symmetrical about a vertical axis and the interior is a cone. It is held so that it can turn freely about a horizontal axis through the centre of gravity, which lies above the vertex of the cone. Shew that as water is poured in, the equilibrium is at first stable, and find the condition that it should subsequently become unstable. Shew in particular that, if the semi-vertical angle of the cone exceeds 30°, the equilibrium always becomes unstable before the vase is full.   [I.]

21. A vessel containing water, whose shape is that of a right circular cone, can swing about a fixed horizontal axis which cuts the axis of the cone at right angles in A. AN is drawn normal to the cone, and NM is drawn perpendicularly to the axis of the cone. Prove that in order that the position of equilibrium may be stable for small displacements, the centre of gravity of the water must be below M, the weight of the vessel being neglected.   [M. T.]

22. The inside of a basin is in the form of the portion of a paraboloid of revolution which is cut off by a plane perpendicular to the axis through the focus of a generating parabola, and the centre of gravity of the basin is at half its depth. The basin is pivoted so that it can swing freely about a diameter of its edge which is horizontal. Shew that the basin may be filled to any extent with water and will remain stable provided the weight of the basin is greater than $\frac{10}{3}$ times the weight of water required to fill it.   [I.]

23. A hollow cylindrical vessel, of radius r, is supported at the ends of a diameter of the cross-section in which is situated the centre of gravity of the vessel. The vessel contains water to a depth h. Prove that the vertical position is unstable unless $r < b\sqrt{2}$ and h lies between $b \pm \sqrt{(b^2 - \frac{1}{2}r^2)}$, where b is the distance of the centre of gravity of the vessel from its base.   [I.]

24. The walls of a ship being vertical in the neighbourhood of the water line, shew that as the draught of water changes on passing from fresh water to salt water, the height of the metacentre above a plane midway between the old and new water lines varies directly as the specific gravity of the water.   [C.]

25. A wall-sided ship with a flat bottom draws a depth d of water when its weight, as loaded, is W; also h is the height of the metacentre above the centre of buoyancy. Prove that, if the ship is to be loaded with an extra cargo W' so that the metacentric height is not to be diminished, then the centre of gravity of the extra cargo must not be at a greater height above that of the ship, as originally loaded, than

$$\frac{1}{2}\left(1 + \frac{W'}{W}\right)d - h.$$ [M. T.]

26. A uniform rod of length a and specific gravity s less than that of water has one end fixed at a point na above a large vessel of water and hangs partially immersed. Prove that the whole potential energy in a

position inclined at an angle θ to the vertical differs from that in the vertical position by a quantity proportional to

$$\{n^2 \sec \theta - (1-s)\}\{1 - \cos \theta\},$$

and find the position of stable equilibrium for different values of the ratio $n^2 : 1 - s$. [M. T.]

27. A log of square section floats in water with the two square faces vertical and three of the edges perpendicular to them wholly immersed. Shew that there are three positions of equilibrium with a given edge not immersed, provided the specific gravity of the substance of the log lies between $\frac{13}{32}$ and $\frac{1}{2}$; and that if this condition be satisfied the two unsymmetrical positions are stable for rolling displacement, and the symmetrical position is unstable. [M. T.]

28. For a long rectangular beam, floating with two of the long edges immersed, prove that the curve of buoyancy, for displacements in which these edges remain horizontal, is an arc of a parabola.

The weight of the beam is W, and its specific gravity is $\frac{1}{2}$, the cross-section is a rectangle of sides $2a$ and $6a$. The beam is floating in water with its shortest edges vertical; prove that if a very small weight kW is placed at the middle point of one of the upper long edges, the beam tilts through an angle $\frac{5}{8}k$ approximately. [I.]

29. A uniform solid cylinder of length l and radius a, which can turn round a diameter of its upper end as a fixed axis, is held vertically immersed in water to a depth h; shew that this vertical position is not stable unless the ratio of the weight of the cylinder to the weight of the water displaced exceeds $2 - \dfrac{h}{l} - \dfrac{a^2}{2hl}$. [I.]

30. A cylindrical object of uniform density and any cross-section having a plane of symmetry has its ends at right angles to its generators and floats in water with its axis vertical. It is found that it will also float in equilibrium with the generators making an angle θ with the vertical, the plane of symmetry remaining vertical. Shew that the centre of gravity of the cylinder must be at a height $k^2(1 + \frac{1}{2}\tan^2 \theta)/h$ above the centre of buoyancy for the vertical position, where k is the radius of gyration of the cross-section of the cylinder about an axis through its centroid normal to the plane of symmetry, and h is the length of the cylinder immersed in the vertical position.

By a consideration of the couples acting for inclinations other than θ, shew that the sloping position is stable and the upright position unstable. [M. T.]

31. In a liquid of depth a a prolate spheroid is half immersed with its axis vertical, so that it rests on the bottom. Prove that equilibrium will be stable if the depth of the centre of gravity below the centre of figure exceeds

$$\left(1 - \frac{5}{8}\frac{W'}{W}\right)\frac{a^2 - b^2}{a},$$

where W is the weight of the spheroid, W' that of the displaced liquid and $2b$ is the length of one of the equal axes of the spheroid. [I.]

ANSWERS

3. Unstable when the diameter of the semi-circular section is vertical.

5. If $2n^2 > 1$, the upright position is stable for all specific gravities.

7. $M . OG/CO$, where M is the mass of the body, G its centre of gravity, C the centre of the cavity and O that of the hemisphere.

10. $\sigma/\rho < 1 - \cos^6 \alpha$.   **14.** The shorter.

15. Stable only when the least principal axis is vertical.

16. For stability the depth of the water put in must exceed $\frac{1}{3}$ the height of the cylinder.

26. If $n^2 > 1 - s$, the vertical position is stable. If $n^2 < 1 - s$, the position given by $\sec^2 \theta = (1-s)/n^2$ is stable. If $n^2 = 1 - s$, the vertical is the only equilibrium position and it is stable.

Chapter VI

EQUILIBRIUM OF FLUIDS IN GIVEN FIELDS OF FORCE

6·1. If every particle of matter in a given region of space is subject to a force, as the result of some such cause as gravitation, the magnitude of the force being proportional to the mass acted upon, then the region may be called a **field of force**. When a particle is placed at any point in such a field it will be acted upon by a force in a definite direction and of magnitude proportional to its mass, though magnitude and direction may vary from point to point. The *force per unit mass* may be called the **intensity of the field**.

A line drawn in a field of force so that the tangent to it at any point is the direction of the force at that point is called a **line of force**.

Through every point of the field which is not a point of zero force there is one definite direction for the force of the field and therefore one line of force passing through the point. The whole field may therefore be regarded as filled with lines of force no two of which cross one another.

For example in the Earth's gravitational field the lines of force in a given locality are vertical lines, or, more strictly, lines directed towards the Earth's centre.

6·11. In this chapter we propose to consider the case of a fluid in equilibrium under the action of any given field of force. It is evident that an arbitrary distribution of force throughout a fluid would not in general maintain equilibrium, but we shall assume that equilibrium exists and then consider to what limitation the field of force must be subject for this to be the case.

We remark at the outset that we are not for the moment concerned with what produces the field of force, but regard it as something 'external'.

6·2. Pressure derivative in terms of force. Let a fluid be in equilibrium under the action of a given system of external forces.

Let p and $p + \delta p$ be the pressures at two neighbouring points P, P', where PP' is a short length δs in an assigned direction.

About PP' as axis describe a cylinder of cross-section α so small that the linear dimensions of the cross-section are small compared to δs. Let the cylinder have plane ends at right angles to PP'. The thrusts on the ends

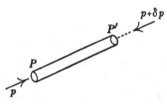

of the cylinder are then, with sufficient accuracy, $p\alpha$ and $(p + \delta p)\,\alpha$.

Again, let ρ denote the mean density of the fluid in the cylinder and let F denote *the component in the direction PP' of the external forces per unit mass of fluid in the cylinder*, so that the external force on the fluid in the cylinder in the direction PP' is $\rho\alpha F \delta s$.

In equilibrium this force must be balanced by the difference of the thrusts of the surrounding fluid on the ends of the cylinder, so that
$$(p + \delta p)\,\alpha - p\alpha = \rho\alpha F \delta s,$$

or
$$\frac{\delta p}{\delta s} = \rho F.$$

Now let $\delta s \to 0$, then ρ becomes the density at P and F becomes the component at P in direction PP' of the external force per unit mass, and we have

$$\frac{\partial p}{\partial s} = \rho F \quad \ldots\ldots\ldots\ldots\ldots\ldots\ldots(1).$$

6·21. Surfaces of equi-pressure. Assuming that the pressure at a point in a fluid can be expressed as a function of the position of the point, e.g. in terms of rectangular coordinates x, y, z, then an equation $p = $ constant represents a surface; and for different values of the constant the equation represents a family of surfaces on each of which the pressure has a constant value. These are **surfaces of equi-pressure**.

From **6·2** (1) it is clear that if p is constant in an assigned direction δs, so that $\partial p/\partial s$ is zero, then the external force has no component in the direction δs. Hence *the direction of the external force at a point must be normal to the surface of equi-pressure through the point*. Conversely, the component F is zero in a direction at right angles to the resultant external force, so that for this direction $\partial p/\partial s = 0$, or p is constant.

We have now discovered the limitation to which a given field of force must be subject if it can maintain a fluid in equilibrium; viz. that it must be possible to draw a family of surfaces which will cut the lines of force at right angles everywhere.

6·22. Pressure gradient. It is obvious that two surfaces of equi-pressure for which p has different constant values cannot intersect.

Let P, P' be points on two neighbouring surfaces of equi-pressure on which the pressures are p and $p + \delta p$. Let $PP' = \delta s$, and let
$$PN \, (= \delta n)$$
be normal to the first surface and meet the second in N. Let the angle $NPP' = \theta$. Then the pressure being the same at N and at P', we have

$$\frac{\partial p}{\partial s} = \lim \frac{\delta p}{\delta s} = \lim \frac{\delta p}{\delta n} \cos \theta = \frac{\partial p}{\partial n} \cos \theta \quad \ldots\ldots\ldots(1).$$

This proves that $\partial p/\partial n$ is a vector whose component in any direction PP' is the space derivative of p in that direction; and this vector $\partial p/\partial n$ is called **the gradient of** p.

The result (1) simply accords with the fact that $\partial p/\partial n$ is the resultant force per unit volume at P and $\partial p/\partial s$ is the component in the direction PP' of the same force.

6·3. The pressure and the potential energy function. If in 6·2 (1) we take δs parallel to the axes of coordinates in turn, we get
$$\frac{\partial p}{\partial x} = \rho X, \quad \frac{\partial p}{\partial y} = \rho Y, \quad \frac{\partial p}{\partial z} = \rho Z \quad \ldots\ldots\ldots\ldots(1),$$

where p, ρ denote the pressure and density and X, Y, Z denote the resolved parts parallel to the axes of the external forces per unit mass at (x, y, z); and ρ, X, Y, Z and p are, in general, functions of x, y, z.

But the total differential dp is given by

$$dp = \frac{\partial p}{\partial x} dx + \frac{\partial p}{\partial y} dy + \frac{\partial p}{\partial z} dz,$$

therefore $dp = \rho \, (X\,dx + Y\,dy + Z\,dz) \,\ldots\ldots\ldots\ldots(2).$

The expression in brackets represents the work which would be done by the external forces on a unit of mass in an infinitesimal displacement from the point (x, y, z). When the forces constitute a conservative system* the expression in brackets is an exact differential, and as it represents positive work done it also represents loss of potential energy and may be denoted by $-dV$, where V is the potential energy function for a unit of mass in the given field. Consequently, when the forces are conservative, (2) may be written

$$dp = -\rho \, dV \,\ldots\ldots\ldots\ldots\ldots\ldots(3).$$

But since dp is an exact differential, therefore $\rho\,dV$ must be an exact differential; so that, if not constant, ρ must be a function of V. Then p is also a function of V, and when any one of these three functions is constant so are the other two. Therefore in a conservative field of force the surfaces of equi-pressure, equi-density and equi-potential energy coincide.

We saw this exemplified in 2·2 and 2·32, for fluid under gravity, where horizontal planes are surfaces of equi-pressure and equi-density and constant potential energy.

6·31. The relation **6·3** (2) is fundamental in the solving of problems in the equilibrium of fluids. It is conveniently remembered in the form 'dp is equal to ρ times the virtual work of the external forces per unit mass'; and it is not necessary to use only Cartesian coordinates, for the virtual work may be expressed equally well by using any set of component forces.

6·311. Example. (i) *Find the surfaces of equi-pressure when a homogeneous liquid is subject to a constant force in a fixed direction (gravity) and a force towards a fixed point and varying as the distance.*

Let r denote distance from the fixed point O and μr the force per unit mass towards O, and let g be the constant force per unit mass. Then, taking the axis of z in the direction of g, **6·3** (2) is in this case

$$dp = \rho \, (g\,dz - \mu r\,dr),$$

* *Statics*, pp. 188, 211.

the sign of the last term being negative because the force μr is opposite to the sense in which r increases.

By integration we get $p = C + g\rho z - \tfrac{1}{2}\mu\rho r^2$,

where C is a constant, or

$$p = C + g\rho z - \tfrac{1}{2}\mu\rho (x^2 + y^2 + z^2).$$

The surfaces $p = \text{const.}$ are clearly concentric spheres.

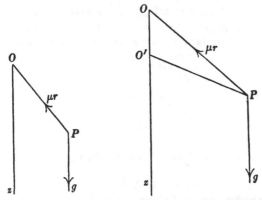

Alternatively, the same result may be obtained by a direct application of **6·21**. Thus, if on Oz we take a point O', such that $\mu \cdot OO' = g$, then O' is a fixed point, and the resultant of the forces μr and g at P, i.e. of forces $\mu \cdot PO$ and $\mu \cdot OO'$, is $\mu \cdot PO'$ in magnitude and direction. Hence at every point of the fluid the external force is directed towards a fixed point, and the surfaces of equi-pressure being orthogonal to these radiating lines are spheres with centre O'.

(ii) *A compressible liquid is at rest under gravity. Defining the compressibility, K, by the relation $(\rho - \rho_0)/\rho_0 = K(p - p_0)$, where ρ and p are the density and pressure respectively, and ρ_0 and p_0 refer to the free surface, and assuming K to be constant, shew that at a depth z below the free surface*

$$\frac{dp}{dz} = g\rho_0 \{1 + K(p - p_0)\}$$

and that $\rho = \rho_0 e^{Kg\rho_0 z}$.

A mine shaft has the same horizontal cross-section at all depths. It is filled with water to a depth h. Shew that if the density of water were everywhere equal to that at the surface, the water would rise in the shaft a distance $\tfrac{1}{2}Kg\rho_0 h^2$. It may be assumed that K is small.   [M. T.]

Measuring z downwards from the free surface, since gravity is the only external force, from **6·3** (2) the pressure is given by

$$dp = \rho g\, dz \quad \dots\dots\dots\dots\dots\dots\dots\dots\dots(1).$$

But $\rho = \rho_0\{1 + K(p - p_0)\} \quad \dots\dots\dots\dots\dots\dots(2),$

therefore $dp/dz = g\rho_0\{1 + K(p - p_0)\} \quad \dots\dots\dots\dots\dots(3).$

Again, from (2) $d\rho = \rho_0 K \, dp,$

therefore from (1) $dp = gK\rho\rho_0 \, dz,$

or $\dfrac{dp}{\rho} = gK\rho_0 \, dz,$

whence by integration $\log \rho = C + gK\rho_0 z,$

and since $\rho = \rho_0$ when $z = 0$, therefore $C = \log \rho_0$ and

$$\log(\rho/\rho_0) = gK\rho_0 z, \quad \text{or} \quad \rho = \rho_0 e^{Kg\rho_0 z} \quad \text{...............(4).}$$

For the second part of the question, let A denote the horizontal section of the mine shaft. Then the mass of water in the shaft

$$= A \int_0^h \rho \, dz,$$

and from (4)

$$= A\rho_0 \int_0^h e^{Kg\rho_0 z} \, dz$$

$$= \frac{A}{Kg} (e^{Kg\rho_0 h} - 1)$$

$$= \frac{A}{Kg} (Kg\rho_0 h + \tfrac{1}{2}K^2 g^2 \rho_0^2 h + \dots)$$

$$= A\rho_0 (h + \tfrac{1}{2}Kg\rho_0 h^2),$$

neglecting higher powers of K than the first.

This shews that if the density of the water were uniform and equal to ρ_0, the total mass remaining the same, the depth of the water would exceed h by $\tfrac{1}{2}Kg\rho_0 h^2$.

6·4. Gravitating matter. According to Newton's law of gravitation two particles of masses m and m' at a distance r apart attract one another with a force $\gamma \dfrac{mm'}{r^2}$, where γ is the gravitation constant, i.e. γ is the force between two particles of unit mass at unit distance apart.[*] It follows that every distribution of matter produces a field of force, which is measured at any point by the resultant force on a particle of unit mass placed at the point. In the *Theory of Attractions* this field of force is calculated for different distributions of matter. For our present purposes, viz. applications to simple problems in hydrostatics, we shall assume some results first proved by Newton, viz.

The attraction of a spherical shell of uniform density at points inside the shell is zero and at points outside the shell is the same as if the whole of the mass were condensed into a particle at the centre.

[*] *Dynamics*, Pt I, p. 167.

Since a uniform solid sphere can be divided into uniform spherical shells, it follows that the attraction of a uniform solid sphere at an external point at a distance r from its centre is $\gamma M/r^2$, where M is the whole mass; or $\frac{4}{3}\gamma\pi\rho a^3/r^2$, if a is the radius and ρ the density.

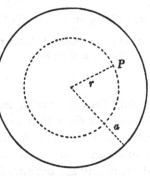

For the attraction at an internal point P at a distance r from the centre, a concentric sphere of radius r may be regarded as dividing the matter into (i) a uniform shell to which P is internal and exerting no attraction at P, and (ii) a uniform sphere of radius r to which P is external and exerting an attraction $\frac{4}{3}\gamma\pi\rho r^3/r^2$ or $\frac{4}{3}\gamma\pi\rho r$ at P; so that the attraction at internal points is $\frac{4}{3}\gamma\pi\rho r$ and is independent of the radius of the sphere.

6·41. Example. *A uniform solid sphere of radius a and density ρ is surrounded by a concentric shell of gravitating liquid of outer radius b and density σ, it is required to find the pressure at a point in the liquid.*

For the field of force in the liquid at a distance r from the centre, we observe that the shell of liquid external to the sphere of radius r exerts

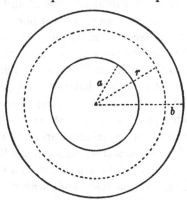

no force, and the matter inside the sphere of radius r attracts as though collected at the centre, i.e. with a force

$$\gamma\{\tfrac{4}{3}\pi\rho a^3 + \tfrac{4}{3}\pi\sigma(r^3 - a^3)\}/r^2$$

or

$$\tfrac{4}{3}\gamma\pi(\rho - \sigma)\frac{a^3}{r^2} + \tfrac{4}{3}\gamma\pi\sigma r.$$

Hence from 6·3 (2) the pressure is given by

$$dp = -\sigma\left\{\tfrac{4}{3}\gamma\pi(\rho-\sigma)\frac{a^3}{r^2} + \tfrac{4}{3}\gamma\pi\sigma r\right\}dr,$$

or, on integration, $p = C + \tfrac{4}{3}\gamma\pi\sigma\left\{(\rho-\sigma)\frac{a^3}{r} - \tfrac{1}{2}\sigma r^2\right\},$

where C is a constant. If the pressure vanishes outside the liquid, then C is determined by the condition that $p = 0$ when $r = b$, so that

$$p = \tfrac{4}{3}\gamma\pi\sigma\left\{(\rho-\sigma)a^3\left(\frac{1}{r}-\frac{1}{b}\right) - \tfrac{1}{2}\sigma(r^2-b^2)\right\}.$$

6·5. Liquid moving in relative equilibrium. Simple considerations shew that when a mass of liquid in motion is in a state of relative equilibrium, i.e. when the whole mass is moving as though it were a solid body, then the problem of determining the pressure is really a statical one.

In the statics of a fluid at rest each element of the fluid is in equilibrium under the action of the external forces and the thrust of the surrounding fluid upon it. Denoting these forces by R and T, in this case R and T balance one another. But when a liquid is in motion in the manner described above each element of mass m has a definite acceleration f, and the 'effective force' mf is the resultant of all the forces acting upon the element, i.e. of the forces R and T. Consequently a force mf *reversed in direction* would balance R and T. Hence the problem may be regarded as a statical one if we regard the liquid as at rest and compound with the external forces R on any element the effective force mf of the element reversed in direction. The resultant thrust T of the surrounding liquid on an element of itself therefore balances the external force R and the reversed effective force mf. And the surface of equipressure through a point is *not* now at right angles to R, as in 6·21 (where T and R were equal and opposite), but it is at right angles to T. This is obvious if we consider the liquid thrust on a short cylinder of liquid whose axis lies in a surface of equipressure. The thrusts on the plane ends are equal and opposite, so that the resultant thrust T must be at right angles to the axis of the cylinder; and this being true for axes in all directions along the surface of equi-pressure, it follows that T is normal to the surface.

We will illustrate this theory by a simple example.

6·51. Example. *A tank of heavy homogeneous liquid is moved horizontally with uniform acceleration f. Find the pressure at any point.*

Let the acceleration of the tank be from left to right in the figure. Then the external force on an element of liquid of mass m is mg vertically downwards and its *reversed* effective force is mf from right to

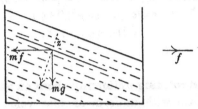

left. The resultant liquid thrust on the element balances the resultant of these two forces and therefore makes an angle $\tan^{-1}(f/g)$ with the vertical.

The surfaces of equi-pressure being at right angles to the resultant liquid thrust at every point are parallel planes making an angle $\tan^{-1}(f/g)$ with the horizontal; and the free surface of the liquid is one of these surfaces of equi-pressure.

We may then find the pressure at a point whose distance from the free surface is z, by drawing the perpendicular z to the free surface and describing a thin cylinder of cross-section α about this perpendicular,

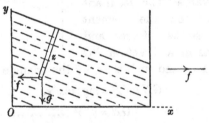

having a plane base parallel to the free surface. Then if p is the required pressure and Π the atmospheric pressure, the difference of the thrusts on the ends of the cylinder is equal to the resultant of the forces f, g per unit mass acting upon it, or

$$p\alpha - \Pi\alpha = \rho z \alpha \sqrt{(f^2 + g^2)},$$

i.e. $$p = \Pi + \rho z \sqrt{(f^2 + g^2)} \quad \text{.......................(1).}$$

Alternatively, we may use the equation **6·3** (2) if we include the *reversed* effective force f per unit mass among the 'external forces'.

Then, taking axes Ox, Oy as in the second figure, we have

$$dp = -\rho(f\,dx + g\,dy)$$

and $$p = C - \rho(fx + gy) \quad \text{.......................(2),}$$

where C is a constant.

Therefore the surfaces of equi-pressure are the planes $fx + gy = $ const., so that, if $fx + gy = c$ represents the free surface on which $p = \Pi$, we have

and, by eliminating C, $\Pi = C - \rho c$,

$$p = \Pi + \rho(c - fx - gy) \quad \dots\dots\dots\dots(3).$$

Also, if z denotes the distance of the point (x, y) from the plane $fx + gy = c$, we have $z = (c - fx - gy)/\sqrt{(f^2 + g^2)}$,

so that $p = \Pi + \rho z \sqrt{(f^2 + g^2)}$,

in agreement with (1).

6·6. Liquid rotating uniformly about an axis. Let a mass of liquid rotate with uniform angular velocity ω about a vertical axis Oz. Then at a point P at a distance r from the axis the acceleration of an element of the liquid is $\omega^2 r$ towards the axis, so that the *reversed* effective force of an element of mass m is $m\omega^2 r$ away from the axis, and the external force is the weight mg vertically downwards. The resultant liquid thrust on the element balances these two forces and therefore acts in a direction PG

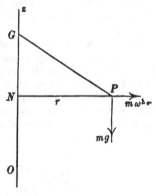

cutting Oz in G, and such that, by the triangle of forces,

$$GN : NP = mg : m\omega^2 r,$$

or $$GN = g/\omega^2 \quad \dots\dots\dots\dots(1).$$

Now by symmetry the surfaces of equi-pressure are surfaces of revolution about the axis Oz and the surfaces are everywhere at right angles to the direction of the resultant liquid thrust. Hence PG is the normal to the surface of equi-pressure through P and PN is an ordinate, and the surface has the property that the *subnormal GN* is constant and equal to g/ω^2. But the only curve having a constant subnormal is a parabola; and its subnormal is half its latus rectum. Hence the surfaces of equi-pressure are paraboloids of revolution, all of the same latus rectum $2g/\omega^2$, and the free surface is one of these paraboloids.

To find the pressure at a point P, draw a vertical line PQ to meet the free surface in Q and about it describe a thin cylinder of horizontal cross-section α. Let M, N be the projections of P, Q on the axis Oz, and let O be the vertex of the free surface. Let p be the pressure at P and Π the atmospheric pressure.

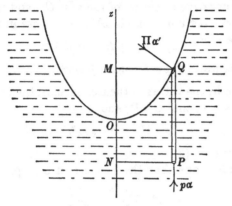

The narrow cylinder meets the free surface in an oblique section of area α', say; and the atmospheric pressure on this is $\Pi\alpha'$ along the normal to the paraboloid. But α is the horizontal projection of α', so this force has a vertical component $\Pi\alpha$. The vertical thrust on the lower end of the cylinder in $p\alpha$, and the difference between the vertical forces on the ends of the cylinder must be equal to its weight; so that

$$p\alpha = \Pi\alpha + g\rho\alpha PQ$$

or $$p = \Pi + g\rho PQ.$$

But $$PQ = NO + OM,$$

and $$OM = MQ^2/(\text{latus rectum})$$

$$= \omega^2 . NP^2/2g,$$

therefore $$p = \Pi + g\rho NO + \tfrac{1}{2}\rho\omega^2 . NP^2 \quad \dots\dots\dots\dots(2),$$

expressing the pressure at P in terms of the vertical and horizontal distances of P from the vertex O of the paraboidal free surface.

It should be noted that if the level of P is above O, the second term in (2) will have a negative sign.

6·61. We obtained the results of **6·6** by elementary considerations—a rather tedious process but instructive as shewing how the separate terms in the expression for p arise. We shall now obtain the same results directly from **6·3 (2)**.

Taking the notation of **6·6**, the external force per unit mass is g in the negative direction of the axis Oz, and the *reversed* effective force per unit mass is $\omega^2 r$ in the sense in which r increases, therefore

$$dp = \rho(-g\,dz + \omega^2 r\,dr)\dots\dots\dots\dots(1);$$

and, by integration, $p = C - g\rho z + \tfrac{1}{2}\rho\omega^2 r^2 \dots\dots\dots\dots(2).$

If the origin be the vertex of the free surface, then $p = \Pi$ when $z = 0$ and $r = 0$, so that $C = \Pi$, and

$$p = \Pi - g\rho z + \tfrac{1}{2}\rho\omega^2 r^2 \dots\dots\dots\dots(3),$$

agreeing with **6·6 (2)**.

This result shews that the surfaces of equi-pressure have equations

$$r^2 = \frac{2g}{\omega^2} z + \text{const.},$$

so that they are paraboloids of revolution of the same latus rectum $2g/\omega^2$.

If the density of the liquid be not constant, equation (1) shews that, since dp is an exact differential, ρ must be a function of $\tfrac{1}{2}\omega^2 r^2 - gz$, so that the same paraboloids of revolution are also the surfaces of equi-density.

6·62. Examples. (i) *Prove that, if a vessel of any form with a plane horizontal lid is just full of liquid of uniform density ρ and the whole rotates with uniform angular velocity ω about a vertical axis, the thrust on the lid is $\tfrac{1}{2}\rho\omega^2 I$, where I denotes the moment of inertia of the area of the lid about the axis of rotation.*

Take the intersection of the axis with the plane of the lid as origin O, let Oz be vertically downwards and let r denote distance from the axis.

Then from **6·3 (2)** $dp = \rho(g\,dz + \omega^2 r\,dr),$

so that $p = C + g\rho z + \tfrac{1}{2}\rho\omega^2 r^2.$

The vessel being 'just full' implies that there is a point at which the pressure vanishes. But the pressure is clearly least when $r = 0$ and $z = 0$, i.e. at the origin. So taking the origin to be the point of zero pressure, we have $C = 0$ and $p = g\rho z + \tfrac{1}{2}\rho\omega^2 r^2.$

Hence, since $z = 0$ on the lid, the pressure at points on the lid is
$$p = \tfrac{1}{2}\rho\omega^2 r^2$$
and the thrust on an element of area dA is $\tfrac{1}{2}\rho\omega^2 r^2 dA$, so that the whole thrust is $\qquad \tfrac{1}{2}\rho\omega^2\Sigma r^2 dA$ or $\tfrac{1}{2}\rho\omega^2 I$.

(ii) *A spherical shell of radius a is just filled with liquid of density* ρ *and the whole rotates with uniform angular velocity* ω *about the vertical diameter. Find at what level the pressure on the shell is greatest and find the resultant thrusts on the upper and lower hemispheres.*

Measure z downwards from the centre of the sphere. Let a be the radius and r denote distance from the vertical diameter. Then

$$dp = g\,dz + \omega^2 r\,dr \quad (6\cdot3\ (2)),$$

so that $p = C + gz + \tfrac{1}{2}\omega^2 r^2$.

It is evident that p is least when $r = 0$ and $z = -a$, i.e. at the top of the sphere. Taking p to be zero at the top, we find that $C = ga$, so that $\qquad p = g(z+a) + \tfrac{1}{2}\omega^2 r^2 \ldots(1)$.

But on the surface of the shell $r^2 = a^2 - z^2$, so that

$$p = g(z+a) + \tfrac{1}{2}\omega^2(a^2 - z^2).$$

Then p has its maximum value when $dp/dz = 0$, and d^2p/dz^2 is negative; i.e. when

$$g - \omega^2 z = 0, \quad \text{or} \quad z = g/\omega^2.$$

To find the thrust on either hemisphere draw the paraboloid of latus rectum $2g/\omega^2$ having its axis vertical and having A as vertex. Then this would be a surface of zero pressure, i.e. a free surface if the rotating liquid were continued in relative equilibrium above the sphere. Hence the thrust on the lower hemisphere CBD is equal to the weight of liquid which would fill the volume $PCBDQ$, between the hemisphere and the paraboloid.

The latus rectum being $2g/\omega^2$, the height of PQ above A is $a^2\omega^2/2g$. Also it is easily proved that the volume of a paraboloid cut off by a plane perpendicular to its axis is half that of the cylinder of the same height on the same base, so the required volume is

$$\tfrac{2}{3}\pi a^3 + \pi a^3 + \frac{1}{4}\frac{\pi a^4\omega^2}{g},$$

and the thrust on the hemisphere CBD is

$$\tfrac{5}{3}g\rho\pi a^3 + \tfrac{1}{4}\rho\pi\omega^2 a^4.$$

The thrust on the upper hemisphere CAD will be less than this by the weight of water in the shell, viz. $\frac{2}{3}g\rho\pi a^3$, and is therefore

$$\frac{2}{3}g\rho\pi a^3 + \frac{1}{4}\rho\pi\omega^2 a^4.$$

(iii) *A small cork of mass m and specific gravity σ is tied by a fine string of length l to a point on the side of a vessel containing water. Prove that when the system is rotating in relative equilibrium with constant angular velocity about a vertical axis, the tension of the string is* $mlg\left(\dfrac{1}{\sigma}-1\right)\Big/h,$ *where h is the height of the cork above the point of attachment.* [C.]

Let P be the cork and A the point of attachment of the string AP. It is evident that the string must exert a force in a plane through the axis of rotation because all the other forces acting on the cork lie in such a plane. Let AP meet the axis of rotation in G and let PN be per-

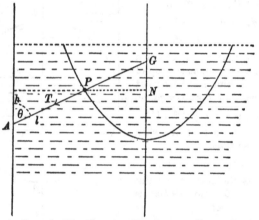

pendicular to the axis. The thrusts of the surrounding water on the cork are the same as they would be on the water displaced by the cork. But its mass is m/σ and it would be in relative equilibrium under the action of its weight mg/σ, the reversed effective force $m\omega^2 NP/\sigma$ and the resultant thrust of the surrounding water. Hence the resultant thrust has components mg/σ vertically upwards and $m\omega^2 PN/\sigma$ towards the axis. Hence the cork is in relative equilibrium under the action of its weight mg, its reversed effective force $m\omega^2 NP$, the foregoing components of thrust and the tension T of the string. Hence T balances an upward vertical force $mg\left(\dfrac{1}{\sigma}-1\right)$, and a horizontal force towards the axis $m\omega^2 PN\left(\dfrac{1}{\sigma}-1\right)$. Hence if the string makes an angle θ with the vertical,

$$T\cos\theta = mg\left(\frac{1}{\sigma}-1\right) \quad\dots\dots\dots\dots\dots\dots(1)$$

and

$$T\sin\theta = m\omega^2 PN\left(\frac{1}{\sigma}-1\right)\dots\dots\dots\dots\dots(2).$$

From (1), we get $T = mlg\left(\dfrac{1}{\sigma}-1\right)\Big/h$ as required; and from (1) and (2)

$\tan\theta = \omega^2 PN/g$, but $\tan\theta = PN/NG$, therefore $NG = \dfrac{g}{\omega^2}$, so that G is the foot of the normal to the surface of equi-pressure through P, or the string is normal to the surface of equi-pressure.

(iv) *A rigid spherical shell is just filled with homogeneous gravitating liquid and the whole rotates with uniform angular velocity ω about a diameter. Find the resultant thrust on the half surface of the shell bounded by the equatorial plane.*

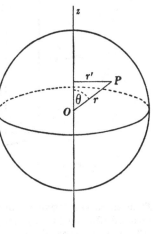

Let a be the internal radius of the shell and ρ the density of the liquid. At a point P at a distance r from the centre O and at a distance r' from the axis of rotation Oz, there is a gravitational force per unit mass $\frac{4}{3}\gamma\pi\rho r$ towards O (6·4), and the reversed effective force per unit mass is $\omega^2 r'$. Therefore from 6·3 (2),

$$dp = \rho\left(-\tfrac{4}{3}\gamma\pi\rho r\,dr + \omega^2 r'dr'\right),$$

and, by integration,

$$p = C - \tfrac{2}{3}\gamma\pi\rho^2 r^2 + \tfrac{1}{2}\rho\omega^2 r'^2 \ldots(1).$$

From this result it appears that the pressure is least when r is greatest and r' least, and this is at the ends of the fixed diameter, where $r' = 0$ and $r = a$. Taking p to be zero at these points, we have $C = \tfrac{2}{3}\gamma\pi\rho^2 a^2$, and therefore

$$p = \tfrac{2}{3}\gamma\pi\rho^2(a^2 - r^2) + \tfrac{1}{2}\rho\omega^2 r'^2 \ldots\ldots\ldots\ldots\ldots(2).$$

At points on the surface of the shell we have

$$p = \tfrac{1}{2}\rho\omega^2 r'^2;$$

indeed we might have foreseen that the gravitational force would contribute nothing to the thrust on the shell.

On the sphere we have $r'^2 = a^2\sin^2\theta$, and resolving parallel to Oz the force on each narrow zone of the upper hemisphere, we have for the resultant thrust

$$\int_0^{\frac{1}{2}\pi} 2\pi a^2\sin\theta\cos\theta\,.\,\tfrac{1}{2}\rho\omega^2 a^2\sin^2\theta\,d\theta = \tfrac{1}{4}\pi\rho\omega^2 a^4.$$

(v) *A cylindrical can, of radius a, with its axis vertical, contains water to height h. It is rotated about its axis and given a horizontal acceleration f. Assuming that the spin is just sufficient not to uncover any point of the base and that the can is sufficiently high for none of the water to spill, shew* *that af is less than both $a^2\omega^2$ and $\sqrt{2}gh$, and that the spin ω satisfies the equation*

$$4gh = a^2\omega^2 + 2f^2/\omega^2. \qquad [\text{I.}]$$

Let the direction of the acceleration f be from right to left in the figure. Take rectangular axes with the origin O at the centre of the base, Oz along the axis of the cylinder and Ox in the direction of f reversed.

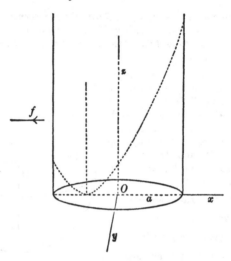

Then there is relative equilibrium under the action of gravity and reversed effective forces f in the direction Ox, and $\omega^2 r$ radially. So that 6·3 (2), in this case, gives

$$dp = \rho\{-g\,dz + f\,dx + \omega^2 r\,dr\},$$

and on integration $\quad \dfrac{p}{\rho} = C - gz + fx + \tfrac{1}{2}\omega^2 r^2,$

or, since $r^2 = x^2 + y^2$,

$$\frac{p}{\rho} = C + \tfrac{1}{2}\omega^2\left\{\left(x + \frac{f}{\omega^2}\right)^2 + y^2 - \frac{2g}{\omega^2}z\right\} - \frac{f^2}{2\omega^2} \ldots\ldots\ldots(1).$$

Since the base is just not uncovered we suppose that the free surface $p = \Pi$ touches the base. This will be so at the point $(-f/\omega^2, 0, 0)$ if $C = \dfrac{f^2}{2\omega^2} + \dfrac{\Pi}{\rho}$, and provided the point lies within the base, i.e. if

$$f < a\omega^2 \qquad\ldots\ldots\ldots\ldots\ldots\ldots(2).$$

Then $\qquad \dfrac{p-\Pi}{\rho} = \tfrac{1}{2}\omega^2\left\{\left(x + \dfrac{f}{\omega^2}\right)^2 + y^2 - \dfrac{2g}{\omega^2}z\right\} \ldots\ldots\ldots(3).$

Hence the equation of the free surface is

$$\left(x + \frac{f}{\omega^2}\right)^2 + y^2 = \frac{2g}{\omega^2}z \qquad\ldots\ldots\ldots\ldots\ldots(4).$$

This is a paraboloid of revolution about an axis parallel to that of the cylinder at a distance f/ω^2 from it.

We now find the equation for ω by expressing the fact that the volume of the cylinder below this paraboloid is the original volume of liquid $\pi a^2 h$.

Hence
$$\pi a^2 h = \iint z\,dx\,dy,$$

where z has the value given by (4), and the integration is over the base of the cylinder.

Therefore
$$\pi a^2 h = \frac{\omega^2}{2g}\iint\left(x^2+\frac{2fx}{\omega^2}+\frac{f^2}{\omega^4}+y^2\right)dx\,dy.$$

But when integrated over a circle of radius a with the origin at the centre
$$\iint x^2 dx\,dy = \iint y^2 dx\,dy = \tfrac14\pi a^4,$$

being moments of inertia about a diameter: also
$$\iint x\,dx\,dy = 0 \quad\text{and}\quad \iint dx\,dy = \pi a^2.$$

Hence
$$\pi a^2 h = \frac{\omega^2}{2g}\pi a^2\left(\frac{a^2}{4}+\frac{f^2}{\omega^4}+\frac{a^2}{4}\right),$$

or
$$4gh = a^2\omega^2 + 2f^2/\omega^2.$$

Since this may be written
$$4gh = \left(a\omega - \frac{\sqrt{2f}}{\omega}\right)^2 + 2\sqrt{2}af,$$

this requires that af should be less than $\sqrt{2gh}$, and we have seen in (2) that af must also be less than $a^2\omega^2$.

EXAMPLES

1. A square lamina is immersed with its plane vertical and its centre of gravity at a given depth in a liquid whose density varies as the depth. Prove that the magnitude of the resultant thrust on the square is independent of the inclination of a side of the square to the surface of the liquid. [I.]

2. A cylindrical bucket, containing water, is raised vertically with a given constant acceleration. Determine the pressure at any point in the liquid.

A cork is held, completely immersed in the bucket, by a string tied to the bottom; find the tension of the string. [I.]

3. A barometer is suspended freely from the roof of a railway carriage which is at rest on a slope of inclination β. If the carriage be now allowed to run freely down the slope, shew that the barometric reading is increased in the ratio $\sec\beta : 1$, but that if the barometer be fixed at right angles to the floor of the carriage, then there will be no alteration in the barometric reading when the carriage moves. [M. T.]

4. A hollow sphere half filled with water is moved with constant acceleration in a horizontal straight line. When the contents are in relative equilibrium, find the pressure at any point.   [C.]

5. A right circular cone with its axis vertical and vertex upwards, just full of heavy fluid of uniform density ρ, is moved horizontally with uniform acceleration f. Shew that the resultant pressure on the base is the greater of the two quantities $\pi g \rho r^2 h$ and $\pi f \rho r^3$, where h is the height of the cone and r the radius of the base.   [M. T.]

6. A circular cylinder of radius r, whose axis is vertical, is filled to a depth h with homogeneous liquid of density ρ. A piston of weight $\pi g \rho \alpha r^2$, which works in the cylinder without friction, is placed on the top of the fluid. Shew that, if the cylinder and liquid rotate about its axis with gradually increasing angular velocity ω, the piston will begin to rise when $\omega r = 2\sqrt{g\alpha}$.   [I.]

7. A sphere of 6 inches radius floats in a rotating fluid of four times its density, and the surface of the fluid meets it along a great circle. Find the angular velocity of rotation.   [I.]

8. A hemispherical shell of radius $2a$ containing water rotates with an angular velocity $\sqrt{2g/3a}$ about its axis which is vertical; a sphere of radius a rests on the water with its lowest point in contact with the shell without pressing on it. The free surface of the fluid passes through the rim of the shell. Prove that

$$\text{density of sphere : density of water} = 3 : 8. \qquad [I.]$$

9. A circular cylinder of radius a floats in liquid at rest with its axis vertical and a length h unimmersed. Shew that, if the cylinder is sufficiently long, it will float in equilibrium with its upper rim in the surface provided that the liquid is made to rotate with angular velocity $2\sqrt{(gh)}/a$.   [M. T.]

10. A hollow circular cylinder of radius r, height h and effective specific gravity σ, floats in water which is rotating round a vertical axis coincident with the axis of the cylinder; and the cylinder is open at its upper end. Shew that the maximum angular velocity ω of the fluid so that the cylinder shall not sink is given by

$$r^2 \omega^2 = 4gh(1-\sigma). \qquad [I.]$$

11. A cylinder of radius a is filled with water to a depth h. A fine vertical tube is at distance d ($>a$) from the axis of the cylinder and communicates with the bottom of the cylinder. If the whole is made to rotate uniformly about the axis of the cylinder, shew that the angular velocity ω necessary to make the water rise in the tube to a height x above the bottom of the cylinder is given by

$$\omega^2 (d^2 - a^2/2) = 2g(x-h). \qquad [I.]$$

12. A tube of small uniform internal section is in the form of an ellipse having its major axis vertical. If the tube be only just filled with liquid whose density at any point varies as the nth power of the distance from the upper focus, shew that the greatest pressure in the liquid exceeds the least by $g\sigma a[(1+e)^{n+1}-(1-e)^{n+1}]/(n+1)e$, where σ is the density at the extremity of the minor axis, e the eccentricity of the ellipse and $2a$ the length of the major axis. [M. T.]

13. A uniform semicircular closed tube of radius r is tightly filled with equal volumes of two fluids of densities ρ and σ respectively which do not mix, and is rotated with angular velocity ω about a vertical radius making an angle α with the line of symmetry. Prove that the pressures at the two ends will be equal if

$$\frac{\omega^2 r}{2g}(\sigma-\rho)=\frac{\sigma}{\cos\alpha+\sin\alpha}-\frac{\rho}{\cos\alpha-\sin\alpha},$$

the fluid of density σ being the lower of the two, and the convexity of the tube being downwards. [M. T.]

14. A liquid whose density is proportional to the depth contains a uniform rod whose density is equal to that of the fluid at a depth equal to half the length of the rod. What are the possible positions of equilibrium? [C.]

15. Prove that, if a circular area of radius a is immersed with its centre at depth h in a liquid, in which the density varies as the depth, the plane of the area being vertical, the centre of pressure is at depth $2a^2h/(a^2+4h^2)$ below the centre of the circle.

A hemisphere, fixed with its plane face vertical, is filled with fluid, in which the density varies as the depth below the highest point of the hemisphere. Shew that the resultant thrust on the curved surface makes an angle $\tan^{-1}(\frac{18}{8})$ with the vertical.

16. A cylinder containing water is rotating round its vertical axis at such a speed that the cylinder is half empty and the bottom is just uncovered. Prove that the volume of liquid left in the cylinder if it rotates uniformly at a higher speed varies inversely as the square of the angular velocity.

17. If a circular cylinder of unit length is filled with a heterogeneous fluid, and the whole rotates in relative equilibrium with given uniform angular velocity about the axis, under the action of no external force, prove that the difference between the pressure at points on the curved surface and that at points on the axis of the cylinder is proportional to the mass of fluid contained. [I.]

18. A rectangular lamina of which the sides are $2a$ and $2b$ is completely immersed in water which is rotating about a vertical axis with uniform angular velocity $(ng/b)^{\frac{1}{2}}$; the plane of the lamina is vertical and

one of its sides ($2b$) touches the surface of the water at its lowest point, the point of contact being the middle point of the side, and the lamina rotates with the same angular velocity as the water. Shew that the depth of the centre of pressure of the lamina is

$$a(8a+6h+nb)/(6a+6h+nb),$$

where h denotes the height of the water barometer.   [M. T.]

19. A fine straight tube of length l, closed at both ends, and inclined to the vertical at an angle α, is just filled with liquid of density ρ. If the tube is rotated with uniform angular velocity ω round a vertical axis through the lower end, prove that the pressure at the highest point is

$$\tfrac{1}{2}\rho\omega^2\sin^2\alpha\,(l-g\cos\alpha/\omega^2\sin^2\alpha)^2,$$

provided that $\omega^2 > g\cos\alpha/l\sin^2\alpha$. What is the pressure when ω^2 is less than this value?   [M. T.]

20. A sphere of radius a floats completely immersed with its centre in the common surface of two liquids which do not mix contained in a cylinder of radius r. Find the angular velocity with which the system must rotate uniformly about the axis of the cylinder so that the radii of the sphere drawn to its intersection with the common surface of the liquids may be inclined to the vertical at an angle of 60°; and shew that the sphere will be depressed through a space $(44r^2 - 39a^2)/54a$, provided that it remains completely immersed.   [I.]

21. A circular cylinder of radius r contains water, and a piston closely fitting the cylinder rests on the surface of the water. The whole is made to rotate with uniform angular velocity ω about the axis of the cylinder, which is vertical. Shew that, provided $\omega^2 r^2 > 4g(b+h)$, the height x, through which the piston rises, is given by the equation

$$\{4g(x-b-h)+\omega^2 r^2\}^2 - 16g\omega^2 r^2 x = 0,$$

where h is the height of the water barometer and b the length of that column of water in the cylinder whose weight is equal to that of the piston.   [M. T.]

22. A rectangular box, full of liquid of density ρ, slides down a rough inclined plane whose coefficient of friction is μ. Shew that the lid, whose weight is w, whose edges are of length a and b and which can turn freely about its lower edge of length b, will not open unless

$$g\rho\mu ab\,(2a\sin\alpha + 3b\cos\alpha) > 6w,$$

where α is the angle which the line of the hinges makes with the line of greatest slope on the plane.   [M. T.]

23. A hollow sphere of radius a, half filled with liquid, is made to rotate with angular velocity about its vertical diameter. If the lowest point of the sphere is just exposed, shew that

$$2g = a\omega^2(2 - \sqrt[3]{4}).$$ [M. T.]

24. A compressible liquid, at rest under gravity, is such that
$$\rho - \rho_0 = \kappa \rho_0 (p - p_0),$$
where κ is a constant, p and ρ are the pressure and density at any point, and p_0 and ρ_0 are their values at the free surface. Prove that, at a depth z below this surface,
$$\rho = \rho_0 e^{\kappa g \rho_0 z}.$$

A right circular cone floats in this liquid with its axis vertical and its vertex at a depth h; prove that, if the density of the liquid were constant and equal to ρ_0, the depth of the vertex would be $h + \frac{1}{12}\kappa g \rho_0 h^2$ approximately, if κ is small. [I.]

25. A thin circular tube of radius a rotates uniformly about a diameter which is vertical: a length s (less than $a\pi$) of the tube is filled with water. Shew how to determine the position of stable equilibrium of the water relative to the tube, when the square of the angular velocity is greater than $\frac{g}{a} \sec \frac{s}{2a}$. [I.]

26. A cylinder of radius b and altitude h floats in a volume Q of fluid at rest in a fixed cylinder of radius a, the axes of both cylinders being vertical. Shew that the distance c between the lowest faces of the two cylinders is given by $\pi a^2 b^2 c = Qb^2 - V(a^2 - b^2)$, where V is the volume of fluid whose weight is equal to that of the floating cylinder.

Given that the cylinders have the same axis, find the angular velocity ω with which the system must rotate about that axis in order that the lowest faces may be infinitely close, and prove that $\omega^2 = \frac{4gc}{a^2 - b^2}$ if $\frac{b^2c}{a^2 - b^2}$ is less than either $\frac{V}{\pi b^2}$ or $h - \frac{V}{\pi b^2}$. [M. T.]

27. A sphere of fluid, whose density ρ at a distance r from its centre varies as $\frac{\sin kr}{kr}$, is self-attracting according to the law of gravitation. Prove that the pressure at distance r from the centre is
$$\frac{2\pi\gamma}{k^2}(\rho^2 - \rho_1^2),$$
where ρ_1 is the surface density of the sphere and γ is the constant of gravitation.

28. A spherical vessel is just filled with heavy liquid, the particles of which attract one another according to the law of gravitation. If the pressure at the highest point vanishes, shew that the resultant thrust across a vertical diametral plane is
$$\pi g \rho a^3 + \frac{1}{3}\pi^2 \rho^2 \gamma a^4,$$
where a is the radius, ρ the density and γ is the constant of gravitation. [I.]

29. A heavy liquid rotates with uniform angular velocity ω about a vertical axis. A solid particle P denser than the liquid is attached by a light string of length l to a point on the axis, so as to be completely immersed and at rest relatively to the liquid. Prove that the string is normal to the surface of equipressure through P, and that if $l > g/\omega^2$, the stable position of relative equilibrium is that in which the string is inclined to the vertical.   [M. T.]

30. A vessel in the form of a right circular cylinder, containing water, is rotated uniformly about a generator, which is vertical, with such an angular velocity that the bottom is just covered. Prove that the rim of the surface of the water is an ellipse. And, if a plane be drawn through the axis of rotation and the axis of the cylinder, prove that the resultant pressure at right angles to this plane, upon the curved surface of the cylinder cut off by this plane, is $\frac{1}{3}wah^2$, where w is the weight of unit volume of the water, a the radius of the cylinder and h the greatest depth of the water.   [M. T.]

31. Prove that, if a mass of homogeneous liquid rotates about an axis and is acted upon by a force to a point in the axis, varying inversely as the square of the distance, the curvatures of the meridian curve of the free surface at the equator and pole are respectively $1/a(1-m)$ and $(1-mb^3/a^3)/b$, where a and b are the equatorial and polar radii, and m is the ratio of the centrifugal force at the equator to the attraction there.   [M. T.]

ANSWERS

2. $\Pi + (g+f)\rho z$, at depth z; where f is the acceleration.
$W(1+f/g)(1-s)/s$, where W is the weight of the cork and s its specific gravity.

4. $\rho z \sqrt{(f^2+g^2)}$, where f is the acceleration and z the distance from the free surface.

7. $2\sqrt{(g/3a)}$, where a is the radius, or $8\sqrt{\frac{2}{3}}$ radian per second.

14. Horizontal or vertical with the middle point of the rod at a depth equal to half its length.

19. Zero.   20. $\sqrt{(88g/27a)}$.

25. The stable position is that in which the centre of the water is at an angular distance $\cos^{-1}\left\{\dfrac{g}{a\omega^2}\sec\dfrac{s}{a}\right\}$ from the vertical diameter.

Chapter VII

PRESSURE OF GASES. THE ATMOSPHERE

7·1. Torricelli's experiment. The barometer. In 1643 Evangelista Torricelli, a pupil of Galileo, performed a simple experiment to demonstrate that the air has weight. A glass tube AB about 3 feet in length, closed at one end and open at the other, is filled with mercury. The tube is then temporarily closed and inverted and reopened with its lower end B below the surface of a dish of mercury. The mercury then sinks in the tube, leaving a vacuum AP at the top, but its upper surface P remains at a height of about 29 inches above the level Q of the mercury in the dish.

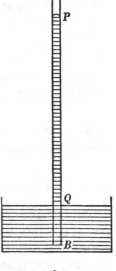

Since, by 2·2, the pressure at the level Q is the same inside and outside the tube, and the pressure inside is due to a depth PQ of mercury, i.e. $g\sigma PQ$, where σ is the density of mercury, therefore there must be an atmospheric pressure of this amount on the free surface of the mercury in the dish, and this pressure is due to the weight of the atmosphere.

Such an apparatus for measuring the pressure of the atmosphere constitutes the simplest form of Barometer.

Other liquids than mercury might be used, but mercury is the most convenient because of its high specific gravity—roughly 13·6. If water were used, the tube would need to be at least 13·6 times as long as for mercury, i.e. about 35 feet in length.

Since the density of mercury and of measuring rods change with the temperature, account must be taken of the temperature at which observations are made in attempting to determine accurately the atmospheric pressure.

**7·2. Relations between pressure, density and tempera-
ture.** The pressure, density and temperature of a given mass of
gas are mutually related, and the relations are expressed by
two laws which can be confirmed by experiment and are found
to be true in a wide range of circumstances.

(i) **Boyle's Law.*** *If the temperature of a given mass of gas
is constant, the pressure varies inversely as the volume*: or

$$pv = \text{const.},$$

where p denotes the pressure and v the volume.

(ii) **Charles's Law.†** *If the pressure of a given mass of gas is
constant, the volume increases by a definite fraction of the volume at
$0°$ C. for every degree centigrade by which the temperature is raised.*

Hence, if v_0 be the volume at $0°$ C. and v the volume at $t°$ C.,
then
$$v = v_0(1 + \alpha t),$$

where α is a definite fraction. For air and most gases α is
approximately $\frac{1}{273}$.

Since the mass is constant, if ρ denotes the density, we have

$$\rho v = \text{constant}.$$

Hence, by combining the laws (i) and (ii), when the pressure,
density and temperature all vary, we have

$$p = k\rho(1 + \alpha t)\dots\dots\dots\dots\dots(1),$$

where k is a constant depending on the nature of the gas.

7·21. It must be remarked that the laws embodied in **7·2**
are not absolute, in the sense of holding good for all ranges of
temperature and pressure. A 'perfect gas' is an ideal substance
for which the gaseous laws are assumed to be true for all
ranges. On this hypothesis the pressure of a given mass of gas
of constant volume would vanish at a temperature $\left(-\frac{1}{\alpha}\right)°$ C.,
i.e. at $-273°$ C.

This temperature is called the **absolute zero** and temperatures measured from this point are called **absolute temperatures**; i.e. the absolute temperature corresponding to $t°$ C. is
$273 + t$.

* Discovered by Robert Boyle in 1662.
† Discovered by J. A. C. Charles in 1787.

We may now write 7·2 (1) in the form

$$p = k\alpha\rho\left(\frac{1}{\alpha}+t\right),$$

or $\qquad\qquad p = R\rho T$ ....(1),

where T is the absolute temperature and R is a constant depending on the nature of the gas.

An equivalent form of the same relation is

$$\frac{pV}{T} = \text{constant} \quad...................(2),$$

where V denotes the volume of the given mass of gas.

7·22. Examples. (i) *The tube of a barometer rises to 34 inches above the mercury in the trough, and the mercury column is 30 inches high. Find what changes are produced in the height of the column by the following operations performed successively:*

(1) *As much air is allowed to rise through the mercury as would, at atmospheric pressure, occupy 2 inches of the tube;*

(2) *A rod of iron whose volume equals that of 5 inches of the tube is allowed to float to the top of the mercury column.*

[*Take specific gravity of mercury to be 13·5, and that of iron to be 7·5.*]
[M. T.]

(1) Let x be the length AC of the tube occupied by air. Then by Boyle's Law the pressure of this air $= 30 \times 2/x$ inches of mercury
$$= 60/x \text{ inches.}$$

But the mercury now stands at a height $CB = 34 - x$ inches; and the pressure of the air in the tube AC plus the pressure due to the column of mercury CB is equal to the atmospheric pressure at B; so that

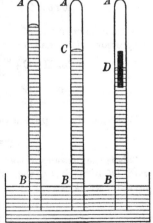

$$\frac{60}{x} + 34 - x = 30,$$

or $\qquad x^2 - 4x - 60 = 0;$

therefore $x = 10$ inches; i.e. the mercury falls to 24 inches.

(2) Let α = cross-section of tube and β = cross-section of iron rod. Then, if l inches is the length of the rod, $l\beta = 5\alpha$. Also the volume of iron immersed in mercury $= \dfrac{7\cdot5}{13\cdot5} l\beta = \tfrac{5}{9}l\beta$; and the volume unimmersed $= \tfrac{4}{9}l\beta = \tfrac{20}{9}\alpha$.

Suppose that the mercury now stands at a level $34 - y$ inches; then the volume occupied by air $=(y-\frac{20}{9})\,\alpha$, and the pressure of the air $=60/(y-\frac{20}{9})$. Then, by equating the pressure at B in the tube to the atmospheric pressure, we have

$$\frac{540}{9y-20}+34-y=30,$$

or
$$540+(9y-20)\,(4-y)=0;$$

which gives $y=10\frac{3}{8}$ approx., so that the mercury now stands at $23\frac{1}{8}$ inches.

(ii) *A bent tube of uniform bore, the arms of which are at right angles, revolves with constant angular velocity ω about the axis of one of its arms, which is vertical and has its extremity immersed in water. Prove that the height to which the water has risen in the vertical arm is*

$$\Pi\,(1-e^{-\omega^2 a^2/2k})/g\rho,$$

a being the length of the horizontal arm, Π the atmospheric pressure, ρ the density of water, and k the ratio of the pressure of the atmosphere to its density.

Reversing the effective force as in 6·61, the equation for the pressure of the air in the horizontal arm AB is, from 6·3 (2),
$$dp=\rho'\,\omega^2 r\,dr,$$

where r denotes distance from the axis of rotation and ρ' the density of air.

But $p=k\rho'$, so that

$$\frac{dp}{p}=\frac{\omega^2 r}{k}\,dr,$$

and this gives on integration

$$\log p=C+\frac{\omega^2 r^2}{2k}.$$

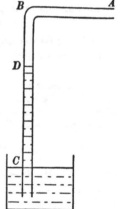

But at the open end of the tube where $r=a$ the pressure p is Π; hence

$$\log\Pi=C+\frac{\omega^2 a^2}{2k},$$

and
$$\log\frac{p}{\Pi}=-\frac{\omega^2}{2k}(a^2-r^2).$$

Consequently the pressure at B, where $r=0$, is given by

$$p=\Pi e^{-\omega^2 a^2/2k}.$$

This is the pressure in the vertical arm, and if there is a column of water DC of height h in the tube, since the pressure at C is atmospheric pressure, we have

$$g\rho h+\Pi e^{-\omega^2 a^2/2k}=\Pi$$

or
$$h=\Pi\,(1-e^{-\omega^2 a^2/2k})/g\rho.$$

7·3. Mixtures of gases. Let two closed vessels of volumes V, V' contain two different gases at the same pressure p and temperature t. If communication is opened between the vessels the gases diffuse into one another and, if no chemical action takes place, they become completely mixed and the mixture has the same pressure p and temperature t as the constituent gases. This experimental fact enables us to deduce that *if two gases at the same temperature when placed separately in turn in a vessel of volume V are at pressures p, p', then when they are placed simultaneously in the same vessel the pressure of the mixture is $p + p'$.*

Consider the gases separately. Let the first have its volume altered so that its pressure changes from p to p'. By Boyle's law its new volume is pV/p'. The two gases are now both at the same pressure p' and have a total volume $V\left(\dfrac{p}{p'} + 1\right)$. They may now be allowed to mix as above without change of total volume or pressure, and the product of the volume and pressure is now $V(p + p')$. If the volume of the mixture is now reduced to V, it follows from Boyle's law that the pressure becomes $p + p'$.

A similar result clearly holds good for a mixture of any number of gases at the same temperature.

7·31. *To prove that, if volumes v_1, v_2 ... v_n of gases at pressures p_1, p_2 ... p_n and absolute temperatures t_1, t_2 ... t_n are mixed without chemical change and V, P, T denote the volume, pressure and absolute temperature of the mixture, then*

$$\frac{VP}{T} = \Sigma \frac{vp}{t}.$$

Making use of **7·21** (2), let us suppose that the volumes and absolute temperatures of all the different gases are changed to V and T; then their corresponding pressures are

$$\frac{T}{V} \cdot \frac{v_1 p_1}{t_1}, \quad \frac{T}{V} \cdot \frac{v_2 p_2}{t_2} \quad ... \quad \frac{T}{V} \cdot \frac{v_n p_n}{t_n}.$$

Now while the temperatures remain T, let the constituent gases be mixed in a single vessel of volume V; then by **7·3** the pressure P of the mixture is the sum of the pressures of the constituents, so that

$$P = \frac{T}{V}\left(\frac{v_1 p_1}{t_1} + \frac{v_2 p_2}{t_2} + ... + \frac{v_n p_n}{t_n}\right),$$

or

$$\frac{VP}{T} = \Sigma \frac{vp}{t}.$$

7·4. Vapours. As stated in **1·14**, by increasing the pressure and decreasing the temperature all gases can be liquefied, but for each gas there is a critical temperature above which no amount of pressure will cause liquefaction.

When the temperature of a gas is above its critical temperature, it is called a permanent gas and it then obeys fairly closely the laws of Boyle and Charles.

When the temperature of a gas is below its critical temperature it is called a vapour; it can then be condensed into a liquid by a sufficient increase of pressure. Vapours are formed by the evaporation of liquids. This may take place by a process so slow as to be imperceptible, or rapidly as when a liquid such as water boils. In this case bubbles of vapour form inside the water, rise to the surface and burst, the pressure of the vapour given off being equal to that of the surrounding atmosphere.

A vapour is said to be **saturated** when in contact with the liquid from which it is given off and in such a state that the slightest increase of pressure or decrease in temperature causes some of the vapour to condense into liquid.

7·41. If, for example, water is contained in a cylinder fitted with a piston, evaporation will take place until the space between the piston and the surface of the water is filled with vapour at a pressure which depends only on the temperature. Evaporation ceases when the vapour is saturated. If, when this stage is reached, the volume is diminished by pushing in the piston and there is no change in temperature, the pressure of the vapour remains unaltered so that some of it must be condensed into liquid.

If the temperature rises, more evaporation will take place and the saturated vapour will have a higher density and pressure.

7·42. Example. *A gas mixed with a saturated vapour is at a pressure p. It is then compressed without change of temperature to 1/nth of its former volume and the pressure is observed to be* p_n. *Determine how the original pressure is divided between the gas and the vapour.*

Let x be the original pressure of the gas. Then $p-x$ is the pressure of the vapour, and this remains unaltered during compression, because at constant temperature the pressure of a saturated vapour is independent of its volume. But owing to compression the pressure of the gas becomes nx, so that $\quad nx + p - x = p_n .$

Hence $\qquad x = \dfrac{p_n - p}{n-1} \quad$ and $\quad p - x = \dfrac{np - p_n}{n-1}$

give the original pressures of the gas and the vapour.

7·5. Specific heats. The capacity of a body for heat may be defined as the quantity of heat required to raise the temperature of the body by one degree.

The **specific heat** of a gas is the quantity of heat required to raise the temperature of a unit of mass of the gas by one degree. But this definition is incomplete until the condition under which the change takes place has been specified. The alternatives are that either the volume or the pressure may be kept constant during the change. In the former case we have *the specific heat at constant volume* and denote it by c_v, and in the latter case we have *the specific heat at constant pressure* and denote it by c_p.

Thus, if dQ is the quantity of heat required to raise the temperature of a unit of mass of gas by a small amount dt *when its volume is kept constant*, then

$$dQ = c_v dt \quad ....(1);$$

or, alternatively, *when its pressure is kept constant*,

$$dQ = c_p dt \quad ....(2).$$

It is evident that c_p is greater than c_v, for the imparting of heat will increase the pressure unless the gas is allowed to expand, so in the second case heat is required not only to raise the temperature but also to expand the gas. This will be proved in the following article.

7·51. Internal energy of a gas. A given mass of gas contains an amount of energy which depends on the motion and configuration of its molecules. The difference between the amounts of energy in two given states of the same mass depends only on those states and not on the mode of passage from one to the other. If E denotes the difference between the amounts of internal energy in an assigned state and in some standard state, then dE is an exact differential of a function determined by the state of the gas.

The state of a mass of gas depends of course on its volume, pressure and temperature, and these are not independent but connected by the relation of 7·21 or some similar 'equation of state'.

The first law of thermodynamics may, for a gas, be expressed in the form
$$dQ = dE + p\,dv \quad\quad\quad\quad (1),$$
i.e. the heat imparted is equal to the increase in internal energy plus the work done in expansion.

Now let us limit our considerations to a perfect gas, for which $pv = RT$ (7·21), and take it as an experimental fact that for such a gas E is a function of T alone.

First let the volume be kept constant while a quantity dQ of heat is imparted. Then, putting $dv = 0$ in (1), we have $dQ = dE$; but, from 7·5 (1),
$$dQ = c_v\,dT,$$
therefore $\quad\quad dE = c_v\,dT, \quad\text{or}\quad c_v = dE/dT.$

Since E is a function of T alone, therefore c_v is a function of T alone. But it is found experimentally that for permanent gases at all but very high and low temperatures c_v is independent of T; therefore c_v is a constant, and (1) may now be written
$$dQ = c_v\,dT + p\,dv \quad\quad\quad (2).$$
But, by hypothesis, $\quad\quad pv = RT,$

so that $\quad\quad\quad\quad p\,dv + v\,dp = R\,dT$

and (2) becomes $\quad dQ = c_v\,dT + R\,dT - v\,dp \quad\quad (3).$

It should be observed that equations (2) and (3) are true generally, irrespective of whether the volume is constant or not. For though we supposed $dv = 0$ in the beginning of the proof, the only purpose of this was to calculate dE/dT which depends on T alone.

Now suppose that the pressure is kept constant while a quantity dQ of heat is imparted. Then by substituting in (3) from 7·5 (2) and putting $dp = 0$, we have
$$c_p\,dT = c_v\,dT + R\,dT,$$
or $\quad\quad\quad\quad c_p - c_v = R \quad\quad\quad (4).$

It follows that for a perfect gas c_p is also a constant and, as stated in 7·5, c_p is greater than c_v.

7·6. Adiabatic changes. When a change of state takes place in a gas without any gain or loss of heat, it is said to be an *adiabatic change*.

In this case, writing $dQ = 0$ in **7·51** (2), we have

$$0 = c_v dT + p\,dv \quad \dots\dots\dots\dots\dots(1).$$

But, for a perfect gas $pv = RT$ and from **7·51** (4)

$$R = c_p - c_v, \quad \text{so that} \quad pv = (c_p - c_v)\,T$$

and
$$p\,dv + v\,dp = (c_p - c_v)\,dT \quad \dots\dots\dots\dots(2).$$

Whence, by eliminating dT from (1) and (2),

$$p\,dv + v\,dp + \left(\frac{c_p}{c_v} - 1\right) p\,dv = 0,$$

or
$$\frac{dp}{p} + \frac{c_p}{c_v}\frac{dv}{v} = 0.$$

This gives, on integration,

$$pv^\gamma = \text{constant} \quad \dots\dots\dots\dots\dots(3),$$

where γ denotes the constant ratio of the specific heats c_p/c_v. The numerical value of this constant is about 1·4, but it varies slightly for different gases.

Relation (3) represents the **law of adiabatic expansion**; i.e. the relation between the pressure and volume of a mass of gas when no heat is imparted or lost.

7·61. Indicator diagrams are graphical representations of the state of a gas in which the ordinates denote pressure and the abscissae denote volume.

On such a diagram, for a perfect gas, *isothermal lines*, or lines of constant temperature, have equations

$$pv = \text{constant};$$

and *adiabatic lines*, or lines of constant quantity of heat, have equations

$$pv^\gamma = \text{constant}.$$

Since $\gamma > 1$, it is easy to see that the adiabatic lines are steeper than the isothermals at their points of intersection.

The adiabatic law is assumed to hold good in all changes which take place so rapidly that there is no time for the transference of heat, as for example in the sudden expansions and condensations of air which constitute sound waves.

7·62. Work done in compressing a gas. Let v denote the volume of gas when its pressure is p, and let there be an external atmospheric pressure Π on the surface of the containing vessel. Let dS denote an element of this surface, then if this element undergoes an *inward* displacement through a normal distance dn, the external work necessary to overcome the opposing pressure is

$$(p - \Pi)\, dS\, dn.$$

And, summing for all elements of the surface which undergo displacement, the work needed is

$$(p - \Pi)\, \Sigma dS\, dn.$$

But $\Sigma dS\, dn$ denotes the decrease in volume due to the displacement of the surface, i.e. $-dv$; so that the work required to reduce the volume by this amount is

$$-(p - \Pi)\, dv,$$

and if the volume is to be reduced from V to V' the work required

$$= -\int_V^{V'} (p - \Pi)\, dv \quad \ldots\ldots\ldots\ldots\ldots\ldots\ldots(1).$$

(i) *Let the change take place isothermally*; then $pv = C$, and the work

$$= -\int_V^{V'} \left(\frac{C}{v} - \Pi \right) dv$$

$$= C \log \frac{V}{V'} - \Pi (V - V')$$

$$= PV \log \frac{V}{V'} - \Pi (V - V') \quad \ldots\ldots\ldots\ldots\ldots(2),$$

where P denotes the initial pressure.

(ii) *Let the change take place adiabatically*; then in (1) $pv^\gamma = C$, and the work

$$= -\int_V^{V'} \left(\frac{C}{v^\gamma} - \Pi \right) dv$$

$$= \frac{C}{\gamma - 1} \left(\frac{1}{V'^{\gamma-1}} - \frac{1}{V^{\gamma-1}} \right) - \Pi (V - V')$$

$$= \frac{PV^\gamma}{\gamma - 1} \left(\frac{1}{V'^{\gamma-1}} - \frac{1}{V^{\gamma-1}} \right) - \Pi (V - V') \ldots\ldots\ldots\ldots(3).$$

If P' denotes the final pressure, then $PV^\gamma = P'V'^\gamma$, so that the work done

$$= \frac{P'V' - PV}{\gamma - 1} - \Pi (V - V') \ldots\ldots\ldots\ldots\ldots(4),$$

and if T, T' are the initial and final absolute temperatures

$$\frac{PV}{T} = \frac{P'V'}{T'} = K,$$

so that the work is also

$$= K \frac{T' - T}{\gamma - 1} - \Pi (V - V') \quad \ldots\ldots\ldots\ldots\ldots(5).$$

It should be noted that the work calculated in case (i) will not be entirely stored up in the internal energy of the gas, because there will be a loss of heat as the volume decreases, as is evident from 7·51 (1).

7·7. Pressure in an isothermal atmosphere. Measuring z vertically upwards, the pressure equation (6·3 (2)) is

$$dp = -g\rho dz \quad \dots \dots (1),$$

where p, ρ denote pressure and density at a height z.

Let p_0, ρ_0 denote the pressure and density at $z = 0$; then in a state of constant temperature

$$p = k\rho \quad \text{and} \quad p_0 = k\rho_0,$$

so that
$$\frac{dp}{p} = -\frac{g}{k}dz$$

or
$$\log p = C - \frac{g}{k}z.$$

But $p = p_0$ when $z = 0$, so that $C = \log p_0$, and

$$p = p_0 e^{-\frac{g}{k}z} \quad \dots \dots (2).$$

If H be the height of a homogeneous atmosphere of density ρ_0 that would produce the pressure p_0, we have

$$p_0 = g\rho_0 H, \quad \text{so that} \quad k = gH,$$

and
$$p = p_0 e^{-z/H} \quad \dots \dots (3).$$

It is evident that this formula also gives an expression for the difference in level z of two stations in terms of the atmospheric pressures at the stations.

7·71. We get a different result if we allow for the difference in the value of the earth's attraction at different heights above its surface. Thus, if g be the measure of gravity at sea-level, and r the earth's radius, the value of gravity at a height z is $gr^2/(r+z)^2$. Hence instead of 7·7 (1) we have

$$dp = -\frac{g\rho r^2 dz}{(r+z)^2} \quad \dots \dots (1),$$

and, with the relation $p = k\rho$ as before,

$$\frac{dp}{p} = -\frac{gr^2 dz}{k(r+z)^2},$$

so that
$$\log p = C + \frac{gr^2}{k(r+z)}.$$

Putting $p = p_0$ when $z = 0$ gives

$$\log p_0 = C + \frac{g}{k}r$$

and
$$\log \frac{p}{p_0} = -\frac{g}{k}\frac{rz}{r+z},$$

or
$$p = p_0 e^{-\frac{g}{k}\frac{rz}{r+z}} \quad\dots\dots\dots\dots\dots(2).$$

7·8. Atmosphere in convective equilibrium. The temperature of the atmosphere is not uniform but diminishes slowly as the height above the earth's surface increases. In an atmosphere at rest conduction would tend to equalize temperature but it would be a very slow process, and in the real atmosphere wind and convection currents are continually changing the state. Lord Kelvin suggested that a state of **convective equilibrium** is a much better working hypothesis than an isothermal state; where by 'convective equilibrium' is implied a state in which, if equal masses of air at any two stations were interchanged, each would assume the pressure, density and temperature of the other; so that the exchange would take place without gain or loss of heat, i.e. adiabatically.

We shall see that this hypothesis implies that the temperature diminishes as the height increases.

In this case the pressure equation is, as before,
$$dp = -g\rho dz \quad\dots\dots\dots\dots\dots(1),$$
where, on the adiabatic hypothesis,
$$p = k\rho^\gamma \quad\dots\dots\dots\dots\dots(2),$$
and, from 7·21,
$$p = R\rho T \quad\dots\dots\dots\dots\dots(3),$$
where T denotes the absolute temperature at height z.

From (1) and (2) $k\gamma\rho^{\gamma-2}d\rho = -g\,dz,$

so that
$$\frac{k\gamma}{\gamma-1}\rho^{\gamma-1} = C - gz,$$

or
$$\frac{\gamma}{\gamma-1}\frac{p}{\rho} = C - gz;$$

and, from (3), $\dfrac{\gamma}{\gamma-1}R(T-T_0) = -gz,$

where T_0 is the absolute temperature at the level from which z is measured.

Thus
$$\frac{T}{T_0} = 1 - \frac{\gamma - 1}{\gamma} \cdot \frac{gz}{RT_0} \quad \dots\dots\dots\dots(4).$$

If H denotes the height of the homogeneous atmosphere,
$$R\rho_0 T_0 = p_0 = g\rho_0 H,$$

and
$$\frac{T}{T_0} = 1 - \frac{\gamma - 1}{\gamma} \cdot \frac{z}{H} \quad \dots\dots\dots\dots(5).$$

If, as in 7·71, we allow for the variation in gravity at different heights and write $gr^2/(r+z)^2$ instead of g in (1), and then proceed as above, we find that

$$\frac{T}{T_0} = 1 - \frac{\gamma - 1}{\gamma} \cdot \frac{rz}{H(r+z)} \quad \dots\dots\dots(6).$$

7·81. Example. *Assuming the temperature of the air to diminish uniformly with the height, prove that the difference of level between two stations is*
$$H \frac{T_0 - T_1}{273} \frac{\log(h_0/h_1)}{\log(T_0/T_1)},$$

where H is the height of the homogeneous atmosphere at $0°$ C., T_0 and T_1 are the absolute temperatures at the stations and h_0, h_1 the barometric heights reduced to $0°$ C. [Consider gravity as constant.] Prove also that, if as an approximation the temperature were taken constant and equal to $\frac{1}{2}(T_0 + T_1)$, the calculated height would be too great by a fraction of its true value equal roughly to
$$\frac{1}{3}\left(\frac{T_0 - T_1}{T_0 + T_1}\right)^2. \qquad \text{[I.]}$$

Let the absolute temperature T at a height z above the lower station be given by
$$T = T_0(1 - cz) \quad \dots\dots\dots\dots\dots(1).$$

The relation between pressure, density and temperature at this level is
$$p = k\rho(1 + \alpha t)$$

or, by 7·21,
$$p = k\rho T/273 \quad \dots\dots\dots\dots(2).$$

The pressure equation $dp = -g\rho dz$

therefore becomes
$$\frac{dp}{p} = -\frac{273g\,dz}{kT} = -\frac{273g\,dz}{kT_0(1-cz)} \quad \dots\dots\dots(3);$$

giving
$$\log \frac{p}{p_0} = \frac{273g}{kT_0 c} \log(1 - cz) \quad \dots\dots\dots\dots(4),$$

where p_0 is the pressure at the lower station.

Hence, if p_1 is the pressure at the upper station and z_1 the difference of level of the stations,
$$T_1 = T_0(1 - cz_1) \quad \dots\dots\dots\dots(5)$$

and
$$\log \frac{p_1}{p_0} = \frac{273g}{kT_0 c} \log(1 - cz_1);$$

or
$$\log \frac{h_1}{h_0} = \frac{273g}{kT_0 c} \log \frac{T_1}{T_0} \quad \dots\dots\dots\dots(6).$$

Also at temperature 0° C.

$$g\rho_0 H = p_0 = k\rho_0, \quad \text{so that} \quad k = gH,$$

and then from (5) and (6), by eliminating c

$$z_1 = \frac{T_0 - T_1}{cT_0} = H\frac{T_0 - T_1}{273}\frac{\log(h_0/h_1)}{\log(T_0/T_1)} \quad\text{............(7)}.$$

Again, if we assume the temperature to be constant and equal to $\frac{1}{2}(T_0 + T_1)$, instead of (2) we have

$$p = \tfrac{1}{2}k\rho(T_0 + T_1)/273$$
$$= k'\rho, \text{ say } \quad\text{.....................(8)};$$

and the pressure equation is

$$dp = -g\rho\,dz$$

or
$$\frac{dp}{p} = -\frac{g\,dz}{k'},$$

so that
$$\log\frac{p}{p_0} = -\frac{gz}{k'},$$

and if z_1' is the calculated height on this hypothesis, then

$$z_1' = \frac{k'}{g}\log\frac{h_0}{h_1}$$

or
$$z_1' = \frac{1}{2}\frac{k}{g}\frac{T_0 + T_1}{273}\log\frac{h_0}{h_1} \quad\text{.....................(9)},$$

where, as before, $k = gH$.

Hence, by comparing (7) and (9),

$$\frac{z_1' - z_1}{z_1} = \frac{1}{2}\frac{T_0 + T_1}{T_0 - T_1}\log\frac{T_0}{T_1} - 1.$$

But
$$\log\frac{T_0}{T_1} = 2\left\{\frac{T_0 - T_1}{T_0 + T_1} + \frac{1}{3}\left(\frac{T_0 - T_1}{T_0 + T_1}\right)^3 + ...\right\},$$

from which it follows that

$$\frac{z_1' - z_1}{z_1} = \frac{1}{3}\left(\frac{T_0 - T_1}{T_0 + T_1}\right)^2,$$

approximately.

EXAMPLES

1. A barometer tube rises to a height of 33 inches above the external mercury surface, and contains as much air as would, at atmospheric pressure, occupy 1 inch of the tube. The true barometric height is known to be 29·6 inches; find the length of tube actually occupied by the air.   [C.]

2. A mass of air occupies 100 cubic inches at 51°C. under a pressure of 56 inches of mercury. What will be its volume at 27° C. under a pressure of 35 inches of mercury?

3. A pressure gauge consists of a U-tube of uniform bore containing mercury, one arm of the tube being closed at the top and containing 15 c.c. of air at atmospheric pressure, and the other arm being connected to the receiver of a condenser. Initially the pressure in the receiver is atmospheric, and by the working of the condenser the volume of air in the gauge is reduced to 3·5 c.c. and the mercury is raised 15 cm. Shew that the pressure in the receiver is then about 4·68 atmospheres, assuming that the height of a mercury barometer is 760 mm. and that the temperature in the gauge is unaltered. [M. T.]

4. Prove that if volumes v_1 and v_2 of atmospheric air are forced into vessels of volumes V_1 and V_2 and a communication is opened between them, a mass of air of volume $(V_1v_2 \sim V_2v_1)/(V_1+V_2)$ at atmospheric pressure will pass from one vessel to the other.

5. Find the atmospheric pressure in pounds per square inch taking the height of the barometer as 30 inches, specific gravity of mercury 13, and the weight of a cubic foot of water as 1000 oz. Also express it in absolute units when a yard, a stone and a minute are the fundamental units. [I.]

6. A U-tube with equal arms of fine uniform bore is placed so that the arms are vertical and the end of one of the arms is closed. The tube is partially filled with mercury, so that there is a length $h/2$ of vacuum above the surface of the mercury in the closed arm, the barometric height being h. Prove that, if it is inclined at an angle of 60° to the vertical, the mercury reaches the top of the closed end; also that if the end that is open when the tube is vertical is closed before inclining it, the length of vacuum becomes $7h/40$ roughly. [I.]

7. The readings of a perfect and a faulty barometer are compared on two occasions and it is found that the differences are x_1 and x_2 inches, the heights in the perfect barometer being h_1 and h_2 respectively. The differences being assumed due to the presence of air in the second barometer, find the correction to a reading h on the latter. No correction for the limited capacity of the reservoir need be made. [M. T.]

8. A thin closed cylindrical vessel, of height a, contains air at atmospheric pressure and floats in water with axis vertical and length b immersed. If a small hole is made in the bottom of the vessel, shew that water leaks in until there is a depth $\frac{ab}{h+b}$ inside, h being the height of the water barometer.

9. A uniform barometer tube $ABCDE$ stands vertically in a vessel of mercury. At the top end, which is closed, there is a space AB filled with air, then a part BC filled with mercury, CD filled with air and then from D downwards there is mercury. If the weights of air in AB and

CD are equal and if the difference of their volumes is one half of BC, shew that the pressure of the air outside is equivalent to a column of mercury whose length is less than the part of the tube which is not immersed by half the length BC.   [I.]

10. A heavy piston can slide without friction in a closed vertical cylinder, and is in equilibrium when there are equal masses m of air in the cylinder above and below the piston and the volumes they occupy are in the ratio $k:1$. Prove that, if the cylinder is inverted, the mass of air which must be forced into the cylinder below the piston to restore the latter to its former position in the cylinder is $2(k-1)m$, the temperature remaining constant.

11. A closed straight tube of fine uniform bore contains mercury in the middle and air at each end. When the tube is vertical the portions occupied by air are of lengths a and b respectively, which become a', b' when the tube is inverted. Prove that the lengths when the tube is horizontal are
$$\frac{aa'(b+b')}{ab+a'b'}, \quad \frac{bb'(a+a')}{ab+a'b'}.$$ [M. T.]

12. A mass M of elastic fluid ($p=k\rho$) surrounds a spherical nucleus of radius a. The fluid is subjected to an attractive force μ/r towards the centre of the sphere. Prove that, if $\mu > 3k$, the fluid may extend to an infinite distance; and that, if ρ_0 is the density at the surface of the sphere,
$$M = 4\pi k\rho_0 a^3/(\mu - 3k).$$

13. A thin uniform circular tube of radius a contains air and rotates with uniform angular velocity ω about an axis in its plane, distant c from the centre; find the pressure at any point, neglecting the weight of the air. If $c < a$, and if p and p' are the greatest and least pressures, prove that
$$\log\frac{p}{p'} = \frac{\omega^2}{2k}(a+c)^2.$$

14. One half of a circular tube, radius a, of fine uniform bore, is occupied by mercury, and the other half by air at atmospheric pressure. The tube is made to rotate about a vertical diameter with uniform angular velocity ω. Shew that, if the air in the tube remains homogeneous, and if $a^2\omega^2 > 2g(a+h)$, h being the height of the mercury barometer, there will be a vacuum of length $2a\phi$ at the bottom of the tube, where $a^2\omega^2 \cos 2\phi = 2gh\pi/(\pi - 2\phi) + 2ga(\sin\phi + \cos\phi)$. [M. T.]

15. A gas satisfying Boyle's law $p = \kappa\rho$ is acted on by forces
$$X = -y/(x^2+y^2), \quad Y = x/(x^2+y^2).$$
Shew that the density varies as $e^{\theta/\kappa}$, where $\tan\theta = y/x$.   [I.]

16. A weightless piston fits into a cylinder of height a with a closed base and vertical generators: initially the cylinder is filled with air at atmospheric pressure and the piston is at the top of the cylinder. If

mercury is poured slowly on the top of the piston, shew that the greatest depth of the upper surface of the mercury below the top of the cylinder is $(\sqrt{a}-\sqrt{h})^2$, where h ($<a$) is the height of the mercury barometer.

What would happen if $h>a$? [M. T.]

17. A box is filled with a heavy gas at a uniform temperature. Prove that, if a is the altitude of the highest point above the lowest, and p, p' are the pressures at these points, the ratio of the pressure to the density at any point is equal to $ag \Big/ \log \dfrac{p'}{p}$. [M. T.]

18. If in the atmosphere p/ρ is constant ($=gk$), calculate on this hypothesis the barometric height on a mountain 6000 feet high, taking 30 inches as the barometric height at sea-level, and k to be 26,000 feet. [C.]

19. The pressure and density of the air are connected by the formula $p = k\rho$. The barometer falls from 30 to 29 inches as a balloon rises 900 feet from the surface; prove that k/g is approximately 26,500 feet. [M. T.]

20. In an atmosphere at rest under gravity, the height H of the equivalent homogeneous atmosphere above any point at height h, measured in terms of air of the density at the height h, is given as a function of h. Shew that the pressure p at height h is given by

$$p/p_0 = e^{-\int_0^h \frac{dh}{H}},$$

where p_0 is the pressure at $h = 0$.

Shew that in an isothermal atmosphere H is a constant independent of h, and deduce that an isothermal atmosphere extends to infinity. [M. T.]

21. The normal height of the barometer at the Riffel Alp 7300 feet above sea-level is 530 mm., that at sea-level being 760 mm.; find the height of the equivalent homogeneous atmosphere. [C.]

22. If the specific gravity of mercury is 13·59 and that of the air at a pressure of 760 mm. of mercury is 0·00129, find the height of the homogeneous atmosphere.

23. The density of mercury is 13·6 and that of air at 760 mm. pressure is 0·001293. Find the reading of the barometer at the top of a building 30 metres high, when the reading at the bottom is 760 mm. [M. T.]

24. If near the earth's surface gravity be assumed to be constant, and the temperature in the atmosphere to be given by $t = t_0 (1 - z/nH)$, where H is the height of the homogeneous atmosphere, shew that the pressure in the atmosphere will be given by the equation

$$p = p_0 (1 - z/nH)^n.$$

Deduce the law of pressure when t is taken constant. [C.]

25. One litre of air at 0° C. and 760 mm. pressure weighs 1·293 grammes. Compute by what fraction of a millimetre the height of a mercury barometer would fall when the instrument is raised through 10 metres, the specific gravity of mercury being 13·6, the temperature of the atmosphere being 0° C. and the height of the barometer at the ground being 760 mm.

Find also the height in an atmosphere everywhere at 0° C. at which the barometer would have fallen to half its height at sea-level. [M. T.]

26. A cubic foot of air at 100 lb. per square inch pressure expands down to atmospheric pressure (14·7 lb. per square inch) following Boyle's law. Shew that the work done is 27,600 ft. lb. approximately.
[M. T.]

27. Two similar barometer tubes A and B stand vertically side by side, rising to a height of 28 inches from large troughs of mercury and water respectively, whose free surfaces are at the same level. The tops of the tubes are connected by a small tube with a stop-cock, which is at first closed. A contains 20 inches of air above a column of mercury, B 10 inches of air above a column of water. Find the alteration in the heights of the columns when the stop-cock is opened; it may be assumed that at the time of the experiment a mercury barometer registers 30 inches, and a water barometer 390 inches.

Discuss also the case in which, before opening the stop-cock, B contained only water, and A 10 inches of air above a column of mercury. [C.]

28. A gaseous atmosphere in equilibrium is such that
$$p = k\rho^\gamma = R\rho T,$$
where p, ρ, T are the pressure, density, and temperature, and k, γ, R are constants. Prove that the temperature decreases upwards at a constant rate α, so that
$$\frac{dT}{dz} = -\alpha = -\frac{g}{R}\frac{\gamma-1}{\gamma}.$$

In a certain atmosphere of uniform composition $T = T_0 - \beta z$, where T_0 and β are constant and $\beta < \alpha$. Find the pressure and density and shew that they both vanish at height T_0/β. [I.]

29. The height of a balloon is calculated from the barometric pressure reading (p) on the assumption that the pressure of the air varies as the density. Shew that if the pressure actually varies as the nth power of the density, there will be an error
$$h_0\left\{\frac{n}{n-1}\left[1-\left(\frac{p}{p_0}\right)^{\frac{n-1}{n}}\right]-\log\frac{p_0}{p}\right\}$$
in the calculated height, where h_0 is the height of the homogeneous atmosphere. [I.]

30. If the absolute temperature T at a height z is a given function $f(z)$ of the height, shew that the ratio of the pressures at two heights z_1 and z_2 is given by
$$\log\frac{p_2}{p_1} = -\frac{g}{k}\int_{z_1}^{z_2}\frac{dz}{f(z)},$$
k being the constant in the equation $p = k\rho T$.

As an aeroplane ascends the temperature and pressure are simultaneously recorded and a curve is drawn plotting the absolute temperature against the logarithm of the pressure. Prove that the height ascended between two readings is
$$-\frac{k}{g}\int_{x_1}^{x_2} T\,dx,$$
where $x = \log p$. [M. T.]

31. Taking into account the variation of gravity with height and assuming that the temperature of the air is constant at all heights, prove that at a height x the pressure p of the air is given by
$$\log\frac{p}{p_0} = -\frac{g_0}{k}\frac{ax}{a(a+x)},$$
where a is the earth's radius, $k = p_0/\rho_0$ and p_0, ρ_0, g_0 are the values of the pressure, density and gravity at the earth's surface. A balloon carrying a self-registering barometer records pressures equivalent to h and h_1 inches of mercury when it ascends to heights equal to fractions α and α_1 of the earth's radius respectively. Prove that
$$\frac{ag_0}{k}\frac{\alpha_1-\alpha}{(1+\alpha)(1+\alpha_1)} = \log\frac{h}{h_1} + 2\log\frac{1+\alpha_1}{1+\alpha}. \qquad\text{[M. T.]}$$

32. A hollow gas-tight sphere containing hydrogen requires a force mg to prevent it from rising when the lowest point touches the ground; the total mass of sphere and hydrogen is M. Shew that the sphere can float in equilibrium with its lowest point at a height h above the ground, where
$$h = \frac{k}{g}\log\frac{M+m}{M},$$
and k is the ratio of the pressure of the atmosphere to its density. [M. T.]

33. On the assumption that the temperature of the atmosphere is constant and equal to $0°$ C., prove that if the barometer readings at two stations are h and h' inches the height of the second station above the first is
$$\frac{5\sigma}{2s\log_{10}e}\log_{10}\frac{h}{h'}\text{ feet,}$$
where σ is the specific gravity of mercury at $0°$ C. and s that of air at $0°$ C. when the mercury barometer stands at 30 inches.

Prove that in order that the atmosphere might be of constant density it would be necessary that the temperature should decrease in arithmetical progression as the height increased in arithmetical progression, the rate being roughly $1°$ C. in 96 feet. [Assume that $\sigma = 13·6$, $s = 0·0013$ and that $\frac{1}{273}$ is the coefficient of expansion for air.] [I.]

34. The pressure at a height z in the atmosphere (assumed to be of constant temperature throughout) is given by $p = p_0 e^{-gz/k}$, where p_0 is the pressure at the earth's surface and k is a certain constant. A large spherical balloon of radius r and total weight W floats with its centre at a height h above the surface of the earth. Shew that h is given by

$$e^{gh/k} = \frac{4\pi p_0 k}{g W} \left\{ r \cosh \frac{gr}{k} - \frac{k}{g} \sinh \frac{gr}{k} \right\}. \qquad \text{[M. T.]}$$

35. Gas expands adiabatically from pressure p_0 and volume v_0 to volume $v_0(1+\alpha)$ and then contracts isothermally to volume v_0. Shew that, if α^3 may be neglected, the work done by the gas is $\frac{1}{2}(\gamma-1)\alpha^2 p_0 v_0$, where γ is the ratio of the specific heats at constant pressure and volume. [External pressure on the containing vessel is ignored.]

$$\text{[M. T.]}$$

36. A gas is compressed to $1/n$th of its original volume, and in the process the relation of pressure to volume is expressed by the equation $pv^\gamma = \text{const.}$, where γ has a constant value. It is then cooled at constant volume to its original temperature. Finally, it is expanded at this constant temperature to its original volume. Shew that the net energy expended on the gas in this cycle is k times the work done in compression, where

$$k = 1 - \frac{(\gamma-1)\log n}{n^{\gamma-1}-1}. \qquad \text{[M. T.]}$$

37. A balloon of given volume, and mean density σ which differs very little from that of the air, is released from the ground where the atmospheric pressure is Π; on the supposition that the mass of air displaced by it in any position is homogeneous and that the absolute temperature at height z is given by $t(z+\beta) = \beta t_0$, shew that the balloon will rise to a height

$$h - \frac{\Pi h^2}{2\beta\left(\Pi - kt_0\, \sigma e^{\frac{gh}{kt_0}}\right)} \text{ (approximately)},$$

where h is the height attained at constant temperature; the formula connecting pressure, density and temperature is $p = k\rho t$ and h/β is small. [I.]

ANSWERS

1. 7·4 inches.   **2.** $148\tfrac{4}{5\cdot7}$ cubic inches.   **5.** 14·1; 418×10^4.

7. $x_1 x_2 (h_1 - x_1 - h_2 + x_2)/\{x_1 (h_1 - x_1 - h) - x_2 (h_2 - x_2 - h)\}$.

13. $\log p = C + \dfrac{\omega^2}{2k} r^2$, where r is distance from the axis.

16. If $h > a$, the piston does not sink.   **18.** 23·82 inches.

21. 6753 yards.   **22.** 8000 metres.   **23.** 757·1 mm.

25. 0·95 mm.; 5540 metres roughly.

27. Mercury drops to about 2 inches and water rises to 26 inches. In the second case water comes over into A and stands at 26 inches in both tubes.

Chapter VIII

HYDROSTATIC MACHINES

8·1. In this chapter we propose to give an account of some of the simpler appliances which depend for their working on the laws of fluid pressure.

8·2. The siphon. This is an appliance for drawing liquid from a stationary vessel without making an aperture. It consists merely of a bent tube with arms of unequal length, which is initially filled with the liquid, by suction or otherwise, and then, without allowing the liquid to run out, placed, as in the figure, with the end A of its shorter arm below the surface of the liquid in the vessel. If the end D is below the level of the free surface of the liquid in the vessel and is then opened, the liquid flows out in a continuous stream. To prove that this should be so, let h, h' denote the heights measured vertically of the top C of the tube above the level B of liquid in the vessel

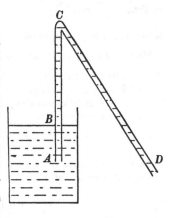

and above D respectively. Then the pressure at B is the atmospheric pressure Π, therefore the pressure in the liquid at C is $\Pi - g\rho h$, and the pressure in the liquid at D is $\Pi - g\rho h + g\rho h'$, where ρ is the density of the liquid. But, by hypothesis, $h' > h$, so that the pressure in the tube at D exceeds the atmospheric pressure and the liquid flows out.

It is clear that the siphon will not work if the height of C above B exceeds the barometric height for the liquid concerned.

8·21. Example. *One end of a siphon is immersed in a liquid whose density at any point is proportional to the nth power of the depth below the surface, the highest point of the siphon being at the level of the free surface of the liquid, and the siphon is filled with homogeneous liquid whose density is equal to that at the immersed end. Prove that liquid will flow out at the free end of the siphon even if it be above the level of the immersed end provided* (i) *that the vertical distance of the free end above the immersed be less than 1/nth of the vertical distance of the former below the surface of the liquid;* and (ii) *that the atmospheric pressure exceeds n/(n+1) of the pressure at the immersed end.* [M. T.]

Measure z downwards from the free surface, let h, h' be the depths of the ends A, B of the siphon, inside and outside the liquid; and let the density be given by

$$\rho = \lambda z^n.$$

The pressure equation is

$$dp = g\rho\, dz = g\lambda z^n dz,$$

so that $p = \dfrac{g\lambda z^{n+1}}{n+1} + \Pi \dots\dots\dots(1),$

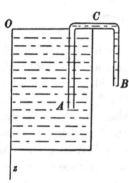

where Π is the atmospheric pressure.

Hence the pressure at $A = \Pi + \dfrac{g\lambda h^{n+1}}{n+1}$. But the liquid in the tube is homogeneous and of density λh^n, so that the pressure in the tube at C is less than the pressure at A by $g\lambda h^{n+1}$; i.e. pressure at C

$$= \Pi - \frac{n}{n+1}\, g\lambda h^{n+1} \dots\dots\dots\dots\dots(2).$$

Also the pressure in the tube at B exceeds that at C by $g\lambda h^n h'$. Hence the pressure at B

$$= \Pi - \frac{n}{n+1}\, g\lambda h^{n+1} + g\lambda h^n h'.$$

For the liquid to flow out this pressure at B must be greater than Π, and this will be so if

$$h' > \frac{n}{n+1}\, h,$$

or if $h - h' < \dfrac{1}{n}\, h'.$

But it is also necessary that the pressure at C shall be positive, i.e. that

$$\Pi > \frac{n}{n+1}\, g\lambda h^{n+1}$$

or $\Pi > \dfrac{n}{n+1}$ (pressure at A).

8·3. The diving bell. This is a cylindrical chamber large enough to hold several persons, closed at the top and open at the bottom. It can be lowered into water by a chain. As it descends the air in the bell becomes compressed but there are generally two tubes communicating with the surface through one of which air can be pumped into the bell while through the other air can escape. The bell is supported partly by the

buoyancy of the water, the amount of which depends on the volume of air in the bell, and partly by the tension of the chain.

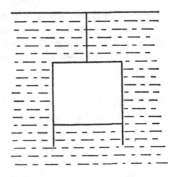

Let H be the height of the water barometer, x the depth of the base of the bell, a its height and y the height to which the water has risen in the bell.

The pressure of the air in the bell is then $H + x - y$ as compared with an initial pressure H, and the volume has been reduced in the ratio a to $a - y$, so that by Boyle's law

$$(a-y)(H+x-y) = aH,$$

an equation which determines y when x is known.

If W denotes the weight in water of the substance of the bell, A its internal cross-section and ρ the density of water, the tension of the chain is

$$W - g\rho A (a-y).$$

If V be the volume of the bell and V' the volume of air at atmospheric pressure H which must be pumped in to clear the bell of water, since in this process a total volume of air $V + V'$ at pressure H is compressed to a volume V at pressure $H + x$, we have

$$(V + V') H = V (H + x)^{\textbf{·}}$$

or

$$V' = Vx/H.$$

From which it follows by differentiating with regard to time, that if the bell descends with uniform velocity v, then air must be pumped in at a rate Vv/H to keep the bell free from water.

8·31. Examples. (i) *A cylindrical diving bell fully immersed is in equilibrium without a chain. Shew that if the exterior atmospheric pressure increases slightly, the ratio of the distance moved through by the bell, if free, to that moved through by the surface of the water in the bell when held fixed is $Hh + x^2 : x^2$ approximately; where H is the height of the water barometer, h the height of the bell, and x the length of that part of it which is filled with air.* [I.]

If y denotes the depth of the base of the bell, then as in **8·3** we have an equation

$$x(H + y - h + x) = hH \quad\quad\quad\quad\quad\dots\dots\dots\dots\dots\dots(1).$$

Firstly, when the bell is floating freely it must displace a constant volume of water, so that x, the length of the part filled with air, must be unaffected by the change of external pressure. Hence, in this case, in (1) we regard x as constant while H and y vary. Differentiating with regard to H, we get

$$x + x\frac{dy}{dH} = h \quad \text{or} \quad \frac{dy}{dH} = \frac{h - x}{x} \quad\dots\dots\dots\dots\dots(2).$$

Secondly, when the bell is held fixed, y is constant and x will vary with H. On this hypothesis, from (1),

$$x + (H + y - h + 2x)\frac{dx}{dH} = h,$$

or, again making use of (1),

$$x + \left(\frac{hH}{x} + x\right)\frac{dx}{dH} = h$$

so that

$$\frac{dx}{dH} = \frac{(h - x)x}{Hh + x^2} \quad\quad\quad\quad\dots\dots\dots\dots\dots\dots\dots(3),$$

and, by comparing (2) and (3), we see that

$$\frac{dy}{dx} = \frac{Hh + x^2}{x^2},$$

which establishes the required result.

(ii) *A cylindrical diving bell whose cross-section is of area A is suspended in water with its flat top at a constant depth d below the surface. A body of volume Aa and specific gravity σ falls off a shelf and floats in the enclosed water. Prove that the depth of the surface of this water below the top of the bell, initially b, will diminish by x, where*

$$x^2 - x(h + b - a + a\sigma) + ha\sigma = 0,$$

the height of the water barometer being denoted by $h - d - b$. Will the bell contain more or less water than before? [C.]

Let $H = h - d - b$ be the height of the water barometer.

Initially the volume of air in the bell is $A(b - a)$, at pressure $H + d + b$, i.e. at pressure h.

When the body falls into the water and floats, its volume immersed is $Aa\sigma$, so that its volume unimmersed is $Aa(1 - \sigma)$. Hence if the water-level in the bell rises a distance x the volume of air in the bell is now

$$A(b - x) - Aa(1 - \sigma),$$

and its pressure is $\quad H + d + b - x \quad \text{or} \quad h - x.$

Hence by Boyle's law

$$(b-x-a+a\sigma)(h-x)=(b-a)h \quad \dots\dots\dots\dots(1)$$

or $$x^2-x(h+b-a+a\sigma)+ha\sigma=0 \quad \dots\dots\dots\dots(2).$$

The original volume of air in the bell is $A(b-a)$ and the final volume is $A(b-x-a+a\sigma).$

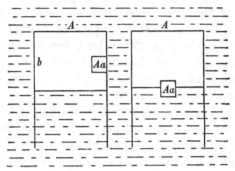

But (1) may be written $(a\sigma-x)h=(b-x-a+a\sigma)x$; and x is positive, since equation (2) has positive roots, and the first factor on the right is positive since when multiplied by A it represents the volume of air in the second case. Therefore $a\sigma$ is greater than x and the volume of air in the bell is increased and the volume of water is decreased.

8·4. The common pump. The common pump consists of a vertical cylinder or barrel AB in which a tightly fitting piston P can be raised or lowered by a lever. The piston contains a valve opening upwards and there is a similar valve at the bottom of the barrel where it communicates with a pipe BC which leads to the water to be raised.

If the action begins with the piston in its lowest position, as the piston rises the air in PB tends to expand and its pressure falls below the pressure in BC, so that the valve at B opens, and as the piston continues to rise the pressure of the atmosphere

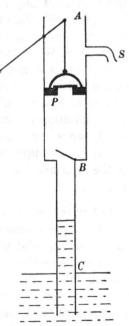

on the water surface at C forces water up the pipe CB. The piston then reaches its highest position and the valve at B closes. The piston then begins to descend. This compresses the air in PB till its pressure exceeds the external atmospheric pressure and opens the valve in the piston, and air escapes from PB as the piston continues to descend. The piston valve then closes and another upstroke follows, the water rising higher in the pipe BC. After a few strokes the water rises into the barrel and then during each downstroke water passes through the piston valve and during each upstroke some escapes through the spout S.

Unless the height BC is less than the height of the water barometer the water will not ascend to B, and for effective working of the pump the height of the whole of the piston range should be less than that of the water barometer.

8·41. The lifting pump. If it is required to lift the water above the pump, the top of the cylinder AB may be closed and instead of the spout S a pipe leading upwards may be connected with the cylinder with a valve opening upwards at the connection. When the water has risen above the piston each upstroke will lift some of it through this valve into the pipe and the only limitation on the height to which the water can be lifted is that it depends upon the strength of the instrument and the operator.

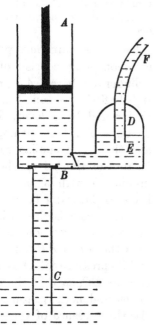

8·42. The force pump. This differs from the common pump in that it has a solid piston and a valve at the bottom of the barrel leading into a chamber D from which emerges a pipe EF, the arrangement being such that after water has risen into the barrel

every downstroke forces some water into the chamber D.
When the water in D has risen to the lower end E of the pipe
some air is imprisoned in the chamber, and this is compressed
as the water rises further and by its pressure tends to maintain a steady flow of water through the pipe EF.

The pipe EF might connect directly with the barrel through
the valve at B, but in that case there would only be an inter-
mittent flow of water during the downstrokes of the piston.

A fire engine is formed of two force pumps communicating
with the same air chamber, so arranged that the upstrokes of
one synchronize with the downstrokes of the other.

8·5. Hawksbee's or the common air-pump. This pump,
for exhausting the air from a vessel C, consists of a vertical
barrel AB in which there is a tightly fitting piston P which
has a valve opening upwards. The vessel C called the receiver
is connected with the barrel by a tube having a valve at B

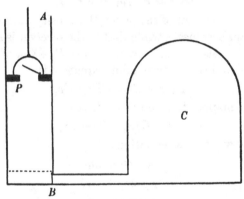

opening into the barrel. The figure shews the valves during a
downstroke of the piston. The air in the barrel is compressed
until the piston valve opens, and then escapes as the piston
descends. During an upstroke the pressure of the air outside
keeps the piston valve closed, and the pressure of the air in the
receiver causes the valve at B to open so that the air in C expands and fills the barrel, and in the next downstroke some of
the air passes through the piston valve and is thus removed
from the apparatus.

The pump ceases to work when the pressure differences on opposite sides of either valve are insufficient to open it.

In general the piston does not descend to the bottom of the barrel to the exclusion of all air below itself so that an upstroke always begins with some air below the piston and this affects the working.

The volume of the barrel below the piston in its lowest position is called a 'clearance'.

Let A denote the volume traversed by the piston, B the volume of the clearance, i.e. the volume of the barrel below the dotted line in the figure, and C the volume of the receiver.

Let the density of the air in the receiver initially be that of the atmosphere, say ρ, and after n complete strokes ρ_n.

At the beginning of the first upstroke the total mass of air below the piston is $(B+C)\rho$, and at the end of the stroke this has expanded to fill a volume $A+B+C$ and the density is ρ_1, so that
$$(A+B+C)\rho_1 = (B+C)\rho.$$

During the descent of the piston the density of the air in the barrel again becomes ρ, while that in the receiver is ρ_1, so that the second upstroke begins with a mass $B\rho + C\rho_1$ of air below the piston, and the corresponding equation is
$$(A+B+C)\rho_2 = B\rho + C\rho_1;$$
and in like manner for the nth upstroke
$$(A+B+C)\rho_n = B\rho + C\rho_{n-1}.$$

We may write these equations
$$\rho_1 = m\rho + k\rho,$$
$$\rho_2 = m\rho + k\rho_1,$$
$$\cdots\cdots\cdots\cdots$$
$$\rho_n = m\rho + k\rho_{n-1},$$
where $m = B/(A+B+C)$ and $k = C/(A+B+C)$.

If we multiply these equations in succession by k^{n-1}, $k^{n-2} \ldots 1$ and add, we find that
$$\rho_n = m\,(1 + k + \ldots + k^{n-1})\,\rho + k^n\rho$$
or
$$\rho_n = \rho\left\{\frac{m\,(1-k^n)}{1-k} + k^n\right\},$$
where m and k have the values stated.

If the clearance B were negligible, the result would be

$$\rho_n = \left(\frac{C}{A+C} \right)^n \rho.$$

It is evident that it is impossible to create a perfect vacuum with this instrument.

8·51. Smeaton's air-pump. This instrument only differs from the common air-pump in that the barrel is closed at the top save for a valve opening outwards. In an upstroke air escapes from the barrel, and in the ensuing downstroke the barrel being closed there is less resistance to the opening of the piston valve from air pressure above it than in the common air-pump.

With the notation of **8·5** suppose that in this case there is an additional clearance of volume B' at the top of the barrel.

Let σ_{n-1} denote the density in the barrel during the downstroke which precedes the nth upstroke. Since the mass of air in the barrel is constant during the downstroke, therefore

$$(A+B+B')\sigma_{n-1} = (A+B)\rho_{n-1} + B'\rho;$$

and since the mass of air below the piston is constant during the next upstroke

$$(A+B+C)\rho_n = B\sigma_{n-1} + C\rho_{n-1}.$$

The elimination of σ_{n-1} between these equations will leave an equation linear in ρ_n, ρ_{n-1} and ρ from which ρ_n can be expressed in terms of ρ as in **8·5**.

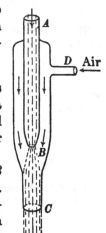

8·52. There are other more effective forms of air-pump, in which air is sucked into a jet of water; based on the hydrodynamical principle that in a steady stream the greater the velocity the smaller the pressure.

Thus water flowing along a tube AB emerges at B from a contracting nozzle. This is surrounded by a jacket which communicates at D with the vessel from which the air is to be exhausted. The contraction of the jet at B increases the velocity of the water and diminishes the pressure in the stream so that air is drawn into the jet and discharged with the water at C.

8·53. Example. *If a Hawksbee's single-barrelled air-pump is perfect except that the piston valve will not open unless the pressure in the barrel exceeds the atmospheric pressure by p, prove that the extra work to be done in completely exhausting the receiver V on account of this defect is equal to that done in compressing a volume V of air at atmospheric pressure until its pressure is increased by p.* [I.]

Let B denote the volume of the barrel, Π the atmospheric pressure and Π_n the pressure in the receiver after n strokes of the piston. Then it is easy to shew that

$$\Pi_n = \left(\frac{V}{V+B}\right)^n \Pi \quad \text{.......................(1)}.$$

During the next downstroke, at any instant before the piston valve opens let P denote the pressure and v the volume of air in the barrel below the piston, so that $Pv = \Pi_n B$.

The piston valve opens as soon as $P = \Pi + p_i$ and then $v = \dfrac{\Pi_n B}{\Pi + p}$.

The work necessary to produce this compression

$$= -\int_B^{\frac{\Pi_n B}{\Pi+p}} (P-\Pi)\,dv = -\int_B^{\frac{\Pi_n B}{\Pi+p}} \left(\frac{\Pi_n B}{v} - \Pi\right) dv$$

$$= -B\Pi_n \log \frac{\Pi_n}{\Pi+p} + B\Pi\left(\frac{\Pi_n}{\Pi+p} - 1\right) \quad \text{.................(2)}.$$

The work done in the corresponding stroke if there were no defect, i.e. if p were zero, would be

$$-B\Pi_n \log \frac{\Pi_n}{\Pi} + B\Pi\left(\frac{\Pi_n}{\Pi} - 1\right) \quad \text{.................(3)},$$

so that the extra work in the nth stroke due to the defect is, by subtraction,

$$B\Pi_n \left\{\log\left(1+\frac{p}{\Pi}\right) - \frac{p}{\Pi+p}\right\},$$

or, from (1), $B\Pi\left(\dfrac{V}{V+B}\right)^n \left\{\log\left(1+\dfrac{p}{\Pi}\right) - \dfrac{p}{\Pi+p}\right\}.$

If we take the sum of like terms for the values 1, 2, 3 ... of n, for an infinite number of terms to represent complete exhaustion, we get

$$V\Pi \left\{\log\left(1+\frac{p}{\Pi}\right) - \frac{p}{\Pi+p}\right\} \text{.....................(4)}.$$

But to compress a volume V of air from pressure Π to pressure $\Pi + p$ requires an amount of work

$$-\int_{V}^{\frac{\Pi V}{\Pi+p}}(P-\Pi)\,dv, \quad \text{where } Pv=\Pi V,$$

$$= -\int_{V}^{\frac{\Pi V}{\Pi+p}}\left(\frac{\Pi V}{v}-\Pi\right)dv$$

$$= V\Pi\left\{\log\left(1+\frac{p}{\Pi}\right)-\frac{p}{\Pi+p}\right\},$$

which is expression (4) and hence the required result.

8·6. The condenser. This is the common bicycle pump, used for increasing the pressure of the air in a given vessel or receiver. The piston which moves in the barrel AB has a valve which opens towards the receiver C and there is a similar valve separating the barrel and the receiver.

Let A denote the volume of the barrel traversed by the piston, B that of the clearance at the bottom of the barrel and C the volume of the receiver.

Then if ρ is the density of the air and ρ_n the density in the receiver after n downstrokes, the original density being ρ, since the mass of air below the piston is constant during a downstroke, we have $\quad (B+C)\rho_n=(A+B)\rho+C\rho_{n-1}$

or say $\qquad\qquad \rho_n=m\rho+k\rho_{n-1},$

where $\qquad m=(A+B)/(B+C) \quad \text{and} \quad k=C/(B+C).$

Similarly $\qquad \rho_{n-1}=m\rho+k\rho_{n-2}$

$$\cdots\cdots\cdots\cdots\cdots$$

$$\rho_1=m\rho+k\rho.$$

Whence, by multiplying the equations in succession by $1, k, k^2 \ldots k^{n-1}$ and adding, we get

$$\rho_n=\rho\left\{m\frac{1-k^n}{1-k}+k^n\right\} \quad\cdots\cdots\cdots\cdots(1),$$

where m and k have the values stated.

8·61. Example. *Shew that, if the piston valve of a condenser does not open until the pressure on the outside exceeds that on the inside by κ times the atmospheric pressure, the number of strokes which can be made before the valve ceases to act is the greatest integer in*

$$1 + \log\{A/(A+B)\,\kappa\}/\log\{(C+B)/C\},$$

where C is the volume of the receiver, A that of the portion of the cylinder traversed by the piston and B that of the portion not traversed by it. [I.]

Let Π be the atmospheric pressure and Π_n the pressure in the receiver after n strokes and let the piston valve cease to act at this stage. At the beginning of the upstroke there is a volume B of air in the clearance at pressure Π_n, and this expands to fill the barrel, so that its volume becomes $A+B$ and its pressure $B\Pi_n/(A+B)$.

Since the piston valve does not open, this pressure plus $\kappa\Pi$ exceeds the external pressure Π, i.e.

$$\Pi_n > \frac{A+B}{B}(1-\kappa)\,\Pi \quad\quad\dots\dots\dots\dots\dots\dots(1).$$

But since the pressures are proportional to the densities, from 8·6 (1),

$$\Pi_n = \Pi\left\{\frac{A+B}{B+C}\cdot\frac{1-\left(\dfrac{C}{B+C}\right)^n}{1-\dfrac{C}{B+C}} + \left(\frac{C}{B+C}\right)^n\right\}$$

$$= \Pi\left\{\frac{A+B}{B} - \frac{A}{B}\left(\frac{C}{B+C}\right)^n\right\}.$$

Hence, from (1),

$$\frac{A+B}{B} - \frac{A}{B}\left(\frac{C}{B+C}\right)^n > \frac{A+B}{B}(1-\kappa)$$

or

$$\left(\frac{C}{B+C}\right)^n < \frac{(A+B)\,\kappa}{A},$$

so that n is the greatest integer in

$$1 + \log\{A/(A+B)\,\kappa\}/\log\{(C+B)/C\}.$$

EXAMPLES

1. The lengths of the arms of a siphon measured vertically are h, k ($h > k$). The siphon is filled with liquid of density σ greater than the density ρ of the liquid in the vessel. Prove that the siphon will begin to work with the end of the longer arm immersed in the liquid in the vessel provided the depth of that end below the surface of the liquid in the vessel exceeds $(h-k)\,\sigma/\rho$.

2. The figure illustrates a siphon for drawing water from a cask. The lengths of pipe DB, BA are 40 and 160 inches respectively. The height of the section B above D is 36 inches. The end A is initially closed and the pipe from A to C filled with water, the air enclosed being at

atmospheric pressure. The end A is then opened and as the water level in CA decreases, the level in DB rises. If a balance is obtained just before the water has risen to B, what length of water will remain in the arm CA?

Shew that the siphon cannot be started by the process described above if the initial length of water in the arm AC is less than 120 inches.

[Assume the height of the water barometer to be 33 feet and neglect the inertia of the water.] [M. T.]

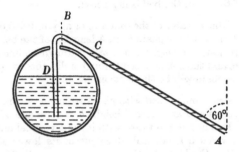

3. Shew that in order to clear a diving bell of water a volume Vc/H of atmospheric air must be pumped in, V being the volume of the bell, c the depth of its lower edge, and H the height of the water barometer. [M. T.]

4. A diving bell is in the form of a cylinder of length a surmounted by a cone of height h. If no air is pumped in when it is immersed, find how far it must be lowered for all the air to be forced into the conical part. Shew that the volume of air at atmospheric density which must now be pumped in that the bell may be filled is $\dfrac{a}{H} + \dfrac{3a}{h} V$, where H is the height of the water barometer, and V is the volume of the bell. [I.]

5. Two cylindrical diving bells, open below and closed at their tops by flat ends, are immersed in water with their closed ends 2 feet above the free surface. The water inside the bells stands at levels 2 and 4 feet below the level outside. Given that, on the establishment of communication between the two bells by means of a small pipe, the water in both attains a common level 3 feet below the level outside, find the ratio of their cross-sectional areas. The height of the water barometer is 33 feet, and the change in the outside water level may be neglected. [M. T.]

6. A diving bell is in the form of a thick circular tube closed at one end. The inner surface of the roof is plane and is at right angles to the axis of the tube which is vertical. The mass of the bell is M and the volume of the metal is V. The cross-section of the inner cylindrical surface is A. The bell is completely immersed in a lake of water of density σ. The height of the water barometer is h. In the initial position of the bell, the depths below the surface of the lake of the inner side of the roof and of the surface of the water in the bell are x_0 and $x_0 + y_0$ respectively, where $x_0 = h$, $y_0 = \frac{1}{2}h$, and the tension of the supporting chain is T_0. In a second position the values are x, $x + y$ and T, where $x = 15h/4$. Find y and T for the second position and shew that

$$T - T_0 = g\sigma Ah/4.$$

The mass of the air in the bell is neglected. [M. T.]

7. The weight in water of the material of a cylindrical diving bell is W, and the tension of the chain is P. A length a of the bell is occupied by air. Shew that, if a weight w of water is drawn up in a bucket into the air space, and the bell is raised through a distance $wa/(W - P)$, the tension of the chain will be the same as before. [M. T.]

8. A cylindrical buoy filled with air at atmospheric pressure has an external radius a and an internal height h. It is made of material of thickness b. It is floating in water with its axis vertical immersed to a depth $h' + b$, when a small leak develops in its lower end. On the assumption that the buoy continues to float, shew that it will sink a further distance x, where $X = x/(a - b)^2$ satisfies

$$a^2 b(2a - b)X^2 - \{(H + h')a^2 + hb(2a - b)\}X + hh' = 0,$$

and H is the height of the water barometer.

Hence shew that, if b is negligible, the condition that the buoy continues to float is $H > h'^2/(h - h')$. [M. T.]

9. If a diving bell descends from the surface with uniform velocity V, shew that the water will ascend a height b in the bell in time

$$\frac{b}{V}\left(1 + \frac{H}{l - b}\right),$$

where l is the length of the bell and H is the height of the water barometer. [I.]

10. A vertical tube which can move freely in the direction of its length is closed at the top and allowed to sink into water. The tube is of thin material, is 40 cm. in length, 1·85 sq. cm. in section, and weighs 50 grammes, and the water is so deep that the tube does not touch the bottom. Find approximately the lengths of the parts into which the tube is divided by the water levels inside and outside the tube, assuming that the height of the water barometer is 10 metres and that a cubic centimetre of water weighs 1 gram. [M. T.]

11. A buoy is formed of a closed hollow circular cone of height a which is weighted so that it floats in stable equilibrium with its axis vertical and vertex downwards. The vertex is at a depth of h feet. Water is now admitted to the inside of the cone through a leak near

the vertex, so that no air escapes, and the cone now floats in equilibrium with its vertex at a depth x, and with a depth y of water inside. If b be the height of the water barometer, shew that

$$x^3 - y^3 = h^3 \quad \text{and} \quad (x - y)(a^3 - y^3) = by^3. \qquad \text{[M. T.]}$$

12. Two cylindrical gasholders, of weights W_1 and W_2 and of cross-sections A_1 and A_2, float in water with their tops at the same height h above the water. The combined quantity of gas in them would occupy a volume V at the atmospheric pressure Π. Prove that

$$h = \frac{\Pi \{V - (W_1 + W_2)/g\rho\} - W_1^2/g\rho A_1 - W_2^2/g\rho A_2}{(A_1 + A_2)\Pi + W_1 + W_2},$$

where ρ is the density of water, and the buoyancy of the water displaced by the sides of the gasholders is neglected. [M. T.]

13. A cylindrical diving bell, whose inner and outer cross-sections are 2 and $2\frac{1}{4}$ square metres respectively, and whose inner and outer heights are 4 and $4\frac{1}{16}$ metres respectively, is forced down into water. If it weighs 6000 kilograms, find the depth at which no force is required to keep it in equilibrium.

Find the force that would have to be applied to keep the bell in this position if 1·9 cubic metres of air at atmospheric pressure were pumped in.

[1 cubic metre of water weighs 1000 kilograms. The height of the water barometer is 10 metres.] [M. T.]

14. A cube of density ρ_0 floats on a liquid of density ρ, and a cylindrical diving bell, of length l and cross-section A, is placed over it and lowered until the cube just touches the top. Shew that the height to which the liquid rises inside the bell is

$$\{A(\rho - \sigma) - \rho a^2\}\{l(\rho - \sigma) - a(\rho - \rho_0)\}/\{\rho(\rho - \sigma)(A - a^2)\},$$

where σ is the density of the atmosphere, and a is the length of an edge of the cube. [M. T.]

15. A body floats on water, the volume of the part not immersed being cA. A cylindrical diving bell of height b and cross-section A is placed over it and then lowered till the top of the bell is at a distance a below the surface of the water. The volume of the floating body which is now not immersed is $(c + \gamma\sigma)A$; shew that γ is the positive root of the quadratic

$$h\gamma^2 + c(h - a - c)\gamma - c^2(a + b) = 0,$$

where h is the height of the water barometer and σ is the specific gravity of the air, σ being small. [M. T.]

16. A canister of weight W without a lid, made of uniform thin material in the form of a right circular cylinder of radius a and height h, is inverted and sunk in water, the whole of the air originally in it being imprisoned. Shew that when the canister is in equilibrium with the

axis vertical the height of the air column is $hH/(H+\sigma h)$, and that the water level inside is at a depth σh below the level outside, where $\sigma = W/\pi a^2 h w$, w being the weight of unit volume of water and H the height of the water barometer.

Shew that the canister is stable in this position only if the centre of gravity is below the centre of gravity of the displaced water, that is, if

$$\frac{a}{2h} < \frac{\sigma h(1+\sigma) - H(1-\sigma)}{H(2-\sigma) - \sigma^2 h}.$$ [M. T.]

17. A glass cylinder contains water of depth h; another glass cylinder of depth k, whose cross-section is e times the cross-section of the first cylinder, is gently lowered, open end foremost, into the water, until it rests on the bottom of the first cylinder, its base not being immersed; and the height of the water in the second cylinder is observed to be y. Prove that the height of the water barometer at the time is

$$(h-y)(k-y)/y(1-e),$$

and that the work done upon the water is

$$\tfrac{1}{2}wA(h-y)^2/(1-e),$$

where A is the area of the cross-section of the second cylinder and w the weight of unit volume of water. [M. T.]

18. A fine tube closed at the upper end stands vertically with its open end in a vessel of mercury (specific gravity σ). Initially the tube contains air at atmospheric pressure; it is then placed in a diving bell which is completely immersed in water. If x be the difference of the water levels inside and outside the bell and h the height of the mercury barometer, find the height y to which the mercury rises in the tube, supposing a length a of the tube to be outside the mercury; and shew that if the mercury in the barometer rises a small distance α the corresponding rise in the tube will be

$$\alpha\left\{1 - \frac{xa}{\sigma y^2}\right\}^{-1}.$$ [I.]

19. A pump consists of a pipe 2 inches in diameter dipping into a well 12 feet below, together with a barrel 6 inches in diameter. When the pipe is full of air at atmospheric pressure, a closely fitting piston is raised 16 inches (from the bottom of the barrel). How far will the water then rise in the pipe?

[The height of the water barometer is about 33 feet.] [C.]

20. Prove that if h, h' be the heights at which the water stands in the lower cylinder of a common pump before and after a stroke, then

$$(h'-h)(h'+h-H-a) + nb(H-h') = 0,$$

where a, b are the lengths of the lower and upper cylinders, n is the ratio of the sectional area of the latter to that of the former, and H is the height of the water barometer. [C.]

21. If in the common pump the heights of the barrel and the pipe leading to it are b, a respectively and n is the ratio of the sectional area of the former to that of the latter, and there is a length c at the bottom of the barrel which the piston does not traverse, shew that if the water just rises into the barrel after two strokes, then

$$h^2\left(\frac{a}{bn}+\frac{2c}{b}-1\right)\left(\frac{a}{bn}+\frac{2c}{b}-2\right)+h\left\{a\left(\frac{3a}{bn}+\frac{6c}{b}-4\right)-n\,(b-c)\right\}$$
$$+a\,(2a+nb)=0,$$

where h is the height of the water barometer. [I.]

22. The barrel of an air-pump is of volume B, but of this a volume b is not traversed by the piston; shew that the density in the receiver never becomes as small as b/B of the atmospheric density. [I.]

23. Air is pumped from a vessel of volume A and forced into a vessel of volume C by means of a Smeaton pump of volume B. Shew that if the whole of the air have originally a density ρ, and if the piston be originally at the end of the cylinder nearest to the vessel A, then after the piston has made $2n-1$ strokes, reaching the end C for the nth time, the density of the air in the two vessels will be

$$\left(\frac{A}{A+B}\right)^n\rho \quad \text{and} \quad \left[A+B+C-\frac{A^n}{(A+B)^{n-1}}\right]\frac{\rho}{C}. \qquad [\text{I.}]$$

24. What must be the pressure of the air in a bicycle tyre, if when the weight borne on it is 150 lb. the area in contact with the ground is 5 square inches?

How many strokes of a bicycle pump must be made to inflate the tyre to this pressure, supposing the air to be initially at a pressure of 14·5 lb. to the square inch, the following dimensions being given: interior diameter of tube 1 inch, diameter of wheel 28 inches, length of stroke of pump 12 inches, internal diameter of pump $\frac{1}{2}$ inch; neglect the interior volume of the connection tube. [C.]

25. If the volume of the receiver of a condenser is A and that of the barrel B, and at the end of each stroke there is a volume C between the piston and the valve, prove that, if ρ be the density of the atmosphere and also the initial density in the receiver, the density after n strokes is

$$\rho\left[1+\frac{B-C}{C}\left\{1-\left(\frac{A}{A+C}\right)^n\right\}\right]. \qquad [\text{I}].$$

26. Air is pumped into a vessel of volume V by means of a compressing pump in which the volume of the barrel varies between v_1 and v_2. The pressure of the air in the vessel at the outset is the atmospheric pressure Π. Assuming the processes to be isothermal, shew that if p_r is the pressure in the vessel after the rth stroke

$$(V+v_1)\,p_{r+1}=Vp_r+v_2\,\Pi.$$

Shew that the pressure of the air in the vessel cannot be raised beyond the value P where $P=v_2\Pi/v_1$, and that

$$p_r=P-\left(\frac{V}{V+v_1}\right)^r(P-\Pi). \qquad [\text{M. T.}]$$

27. A condenser, whose clearance may be neglected, and its receiver are contained in an air-tight room. The air in the room and in the receiver is initially of density ρ; prove that after n strokes the density of the air in the receiver becomes

$$\rho\left[1 + \frac{V+u}{v} - \frac{V+u}{v}\left(\frac{V}{V+u}\right)^n\right],$$

where u is the volume of the barrel of the condenser, v that of the receiver and $V+u+v$ that of the room.   [M. T.]

28. If air be taken from a vessel of volume A and condensed in a vessel of volume A' by a piston working in a cylinder of volume B of which the clearances C and C' at the ends are not traversed by the piston, shew that, neglecting the effect of the weights of the several valves, the limiting ratio of the pressure in A to that in A' is

$$\frac{CC'}{(B-C)(B-C')}$$

and find the limiting pressures in terms of the original atmospheric pressure throughout the whole system.   [I.]

29. In a condenser the sectional area of the piston is A sq. inches, the length of the barrel is a inches, and the volume of the receiver is bA cubic inches. If the piston moves uniformly, traversing the barrel in one second, prove that the rate of working after $2n+t$ seconds $(0 < t < 1)$ from the beginning of the first stroke is $\dfrac{A\Pi a^2}{12}\left(\dfrac{n+t}{b+a-at}\right)$ or $\dfrac{A\Pi a}{12}\cdot\dfrac{t}{1-t}$ ft. lb. per second, according as the receiver valve is or is not open; where Π is the atmospheric pressure in pounds-weight per square inch, and the first stroke begins with the piston fully drawn out.   [I.]

ANSWERS

2. 72 inches.   **4.** Depth of bottom of bell $= a\,(1+3H/h)$.

5. 21 : 20.   **6.** $y = \frac{1}{4}h$; $\ T = g\,(M - \sigma V - \frac{1}{4}\sigma A h)$.

10. 1, 27 and 12 cm. approx.

13. The bottom of the bell is at a depth of 7·5 metres; 1000 kilograms.

18. y is given by $\sigma y^2 - y\,(\sigma h + x + \sigma a) + ax = 0$.

19. 8·1 ft.   **24.** 30 lb. per square inch; 61.

28. $\dfrac{p}{CC'} = \dfrac{p'}{(B-C)(B-C')} = \dfrac{\Pi\,(A+A'+B)}{ACC'+A'(B-C)(B-C')+BC'(B-C')}$

give the ratios of the limiting pressures p, p' to the original pressure Π.

Chapter IX

CAPILLARITY

9·1. There are many familiar phenomena which appear to contradict the theorem of **2·2** that the pressure in a heavy liquid at rest is the same at all points of a horizontal plane, with its corollary that the free surface of such a liquid is a horizontal plane. Among such are (i) the existence of separate drops of liquid, such as drops of water or mercury resting on a table; (ii) the curved surface of water at rest in a tank particularly near the walls of the tank; (iii) the rise of water in a fine tube inserted vertically in a vessel of water, the water in the tube standing at a higher level than the water in the vessel and having a surface concave upwards; (iv) the depression of mercury in a fine tube inserted vertically in a vessel of mercury, the mercury in the tube standing at a lower level than the mercury in the vessel and having a surface convex upwards. The narrower the tubes in (iii) and (iv) the greater is the elevation or depression. Tubes whose bore is of the diameter of a hair, or say 1 mm., are called *capillary tubes* and in them the phenomenon is most marked.

The explanation of these and other kindred phenomena is to be found in the existence of forces of cohesion between the molecules of fluids, the range through which a molecule exerts such a force being exceedingly small—probably of the order of a millionth of a centimetre. It follows that throughout the interior of a mass of homogeneous fluid all molecules are similarly situated as regards the cohesive forces exerted on them by surrounding molecules; but at a boundary surface between two fluids there is in each fluid a very thin layer in which the conditions are not the same as in the interiors, because molecules in either of these layers fall within the range of action of the cohesive forces exerted by molecules in the other layer. In consequence there is associated with every such surface of separation of two fluids or of a fluid and a solid a definite amount of energy; and because the range of the molecular action is so small this energy is independent of the curvature of the surface and depends only on the area and on the nature of the two fluids or the fluid and solid in contact.

9·11. Surface tension. The phenomena are equally well accounted for by the hypothesis of the existence of surface tension. We assume that in the surface of separation of two fluids or of a fluid and a solid there is a uniform tension; i.e. that the stress across any short line δs in the surface is a tension $T\delta s$ at right angles to the line, where T is a constant depending only on the fluids or fluid and solid in contact.

It is easy to see that if such a tension exists it must be uniform. For if ABC be any triangular element of area on the surface and we assume that it is kept at rest by forces T_1, T_2, T_3 acting at right angles to the sides through their middle point, we have by Lami's theorem

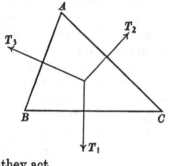

$$T_1 : T_2 : T_3 = \sin A : \sin B : \sin C$$
$$= a : b : c,$$

so that the stresses are all the same constant multiple of the lengths of the lines across which they act.

It can be shewn* by the principle of minimum potential energy that the hypothesis of surface energy proportional to area is equivalent to the hypothesis of a uniform surface tension.

9·2. Angle of contact. Let the surface of separation of two fluids A and B (e.g. air and water) meet the surface of a solid C at an angle α. There are in all three surfaces of separation, viz. between A and B, B and C, C and A. In each of these surfaces there is a tension T_{AB}, T_{BC}, T_{CA} depending on the substances separated; and they must be such that

$$T_{CA} - T_{BC} = T_{AB} \cos \alpha,$$

* Besant and Ramsey, *Treatise on Hydromechanics*, Part I, *Hydrostatics*, §§ 101–103.

so that the *angle of contact* α is constant for the same set of substances.

When A is air, B water and C glass, the angle α is an acute angle; but when A is air, B mercury and C glass, then α is an obtuse angle.

9·21. When a drop of oil is placed upon water it rapidly spreads over the surface in a thin film. This is because the surface tension between water and air exceeds the sum of the tensions between oil and air and between oil and water.

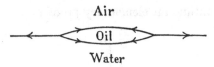

Air

Oil

Water

9·3. Rise of liquid between vertical parallel plates. Let T be the surface tension and α the angle of contact, d the distance between the plates and h the mean rise. Since the atmospheric pressure is the same at the surface of the liquid between the plates as at the external surface of the liquid, it follows that the weight of the elevated liquid is supported by the vertical component of the tensions on the upper boundary. Hence

$$2T\cos\alpha = g\rho hd,$$

where ρ is the density of the liquid.

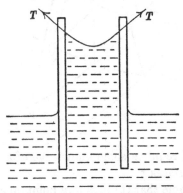

9·31. Rise of liquid in a vertical circular tube. If the figure of **9·3** represents an axial section of the tube and r denotes the radius of the tube, then by similar reasoning

$$2\pi r T\cos\alpha = g\rho\pi r^2 h$$

or $\qquad\qquad\qquad 2T\cos\alpha = g\rho r h.$

It must be observed that these results only refer to a 'mean rise' h, and do not touch the question of the form of the curved surface, which we will consider later.

9·4. Relation between surface tension and pressure. There is a relation between the tension in the surface of separation between two fluids, the difference of the pressures on opposite sides of the surface and the curvature of the surface.

We shall outline an elementary proof of this relation. A

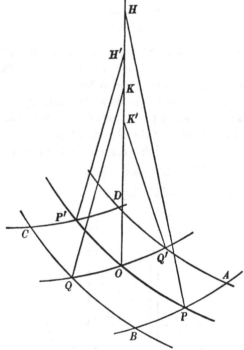

complete proof requires some knowledge of analysis and the geometry of surfaces.*

We shall assume that through any point on a surface, there are two 'lines of curvature' cutting one another at right

* See Besant and Ramsey, *Treatise on Hydromechanics*, Part I, *Hydrostatics*, § 101.

angles, a line of curvature being a line such that at any two points on it sufficiently close together the normals to the surface intersect.

Let $P'OP$, $Q'OQ$ be small elements of the lines of curvature through any point O on the surface of separation of two fluids in which there is a surface tension T.

Take $P'O = OP$ and $Q'O = OQ$, and let the normals to the surface at P, P', Q, Q' intersect the normal at O in H, H', K, K'.

Let planes through the normals PH, QK, $P'H'$, $Q'K'$ at right angles respectively to OP, OQ, OP', OQ' cut the surface in the sides AB, BC, CD, DA of a curvilinear quadrilateral. Let p denote the mean value over this quadrilateral of the excess of the pressure on its concave side over the pressure on its convex side. This pressure difference gives rise to a force $p.AB.BC$ directed along the normal HO, approximately. This force must be balanced by the surface tensions across the sides AB, BC, CD, DA of the elementary area.

But the stress across AB is $T.AB$ at right angles to AB through its middle point and in the tangent plane to the surface, i.e. approximately along the tangent at P to the curve OP. And the component of this force in the direction of the normal OH is $T.AB.OP/OH$. Hence we get an equation

$$T.AB.\frac{OP}{OH} + T.BC.\frac{OQ}{OK} + T.CD.\frac{OP'}{OH'}$$
$$+ T.DA.\frac{OQ'}{OK'} = p.AB.BC.$$

Now let P, P', Q, Q' move up to O, then H, H' move up to coincidence with the centre of curvature of the normal section through $P'P$, so that OH, OH' become the radius of curvature r_1 of this section, and similarly OK, OK' become the radius of curvature r_2 of the perpendicular normal section. These are called the principal radii of curvature of the surface at O. Also p becomes the difference of the pressure on opposite sides of the surface at O, and as the curvilinear quadrilateral shrinks its opposite sides tend to equality, so that, dividing by $AB.BC$, we get

$$T\left(\frac{1}{r_1} + \frac{1}{r_2}\right) = p \quad \dots\dots\dots\dots\dots(1).$$

9·41. On *a sphere of radius r* all normal sections are lines of curvature and all principal radii of curvature are of the same length r, so that the formula of **9·4** becomes

$$\frac{2T}{r} = p,$$

where p denotes the excess of the internal pressure over the external pressure.

On *a surface of revolution* obtained by revolving a curve AB about an axis Ox in the same plane, it is evident that the normals to the surface at all points of a meridian curve AB intersect, so that the meridian

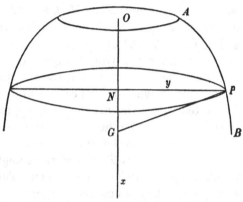

curves are lines of curvature, and the other set of lines of curvature are the circles in planes at right angles to the axis. Also if the normal to the surface at a point P meets the axis in G, then one principal radius of curvature at P is the radius of curvature of the meridian curve APB, and the other is the normal PG, because as the figure rotates round Ox, G will be the intersection of normals at neighbouring points on a circular section, so in this case the formula of **9·4** becomes

$$T\left(\frac{1}{R} + \frac{1}{PG}\right) = p,$$

where R is the radius of curvature of the meridian curve.

9·5. The capillary curve. Let liquid rise to meet a vertical wall AB at an angle α. Let the figure represent a vertical section at right angles to the wall, and let P be a point on the curved surface. Take an axis Ox at right angles to the wall in the natural level of the liquid surface, i.e. the level at which the pressure in this liquid is equal to the atmospheric pressure Π.

Let R be the radius of curvature at the point P (x, y) on the capillary curve, this is one principal radius of curvature of the surface of the liquid, and the surface being cylindrical (with horizontal generators) the other radius of curvature is infinite. Hence from 9·4 (1)

$$\frac{T}{R} = \Pi - p = g\rho y \quad \dots\dots\dots\dots\dots(1),$$

where T is the surface tension, p the pressure at P in the liquid, and ρ the density of the liquid. If we put $4T = g\rho c^2$, we get

$$Ry = \tfrac{1}{4}c^2 \quad \dots\dots\dots\dots\dots\dots(2).$$

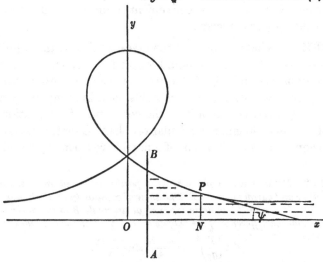

Let ψ denote the angle between the tangent at P and Ox, as in the figure, and let the arc s be measured from the wall. Then $R = -ds/d\psi$, and $ds/dy = -\operatorname{cosec}\psi$, so that

$$y\,dy = \tfrac{1}{4}c^2 \sin\psi\,d\psi \quad \dots\dots\dots\dots\dots(3),$$

and, since y and ψ vanish together,

$$y^2 = \tfrac{1}{2}c^2(1 - \cos\psi) = c^2 \sin^2 \tfrac{1}{2}\psi,$$

or

$$y = \pm c \sin\tfrac{1}{2}\psi \quad \dots\dots\dots\dots\dots(4),$$

where, in the case under consideration the upper sign must be taken.

Again, since $dy/dx = -\tan\psi$,

therefore
$$dx = -\tfrac{1}{2}c\cos\tfrac{1}{2}\psi \cot\psi\,d\psi$$
$$= -\tfrac{1}{4}c(\operatorname{cosec}\tfrac{1}{2}\psi - 2\sin\tfrac{1}{2}\psi)\,d\psi,$$

and
$$x = \tfrac{1}{2}c\log\cot\tfrac{1}{4}\psi - c\cos\tfrac{1}{2}\psi \quad \dots\dots\dots(5),$$

provided the origin be so chosen that $x = 0$ when $\psi = \pi$.

Equations (4) and (5) represent the capillary curve parametrically. It has a loop as in the figure and is asymptotic to the axis of x.

The height above the natural level at which the liquid meets the wall is given by putting $\psi = \frac{1}{2}\pi - \alpha$ in (4), i.e.

$$y = c\sin(\tfrac{1}{4}\pi - \tfrac{1}{2}\alpha) \quad\dotsc\dotsc\dotsc\dotsc(6).$$

In the case of a liquid such as mercury for which the angle of contact is obtuse, it is convenient to measure y downwards and the figure is inverted.

9·51. We have associated the investigation of the capillary curve with the rise of liquid against a vertical wall, but equations (4) and (5) of **9·5** are the result of integrating a differential equation (2) which holds good in many cases, and it is only necessary to adjust the constants of integration to suit the conditions of a particular problem. Equation (2) is also the equation of equilibrium of a flexible rod bent by terminal forces.

9·52. Example. *A plane plate is partly immersed in a liquid of density ρ and surface tension T. The angle of capillarity for the liquid and the plate is α and the plate is inclined at an angle β to the horizontal. Prove that the difference of the heights of the liquid on the two sides of the plate is*

$$4\left(\frac{T}{g\rho}\right)^{\frac{1}{2}}\cos\frac{\pi-2\alpha}{4}\sin\frac{\pi-2\beta}{4}.$$

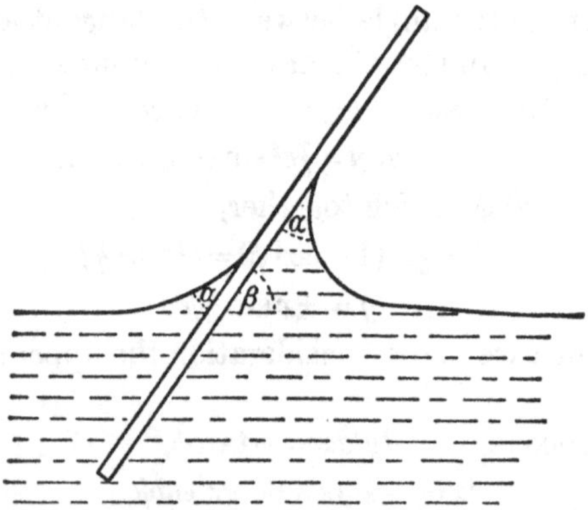

The ordinate of a point on the capillary curve is given by

$$y = c\sin\tfrac{1}{2}\psi \quad (\mathbf{9·5}\ (4)),$$

where ψ is the acute angle which the tangent to the curve makes with the horizontal.

Hence the required difference

$$c \sin \tfrac{1}{2}(\pi - \alpha - \beta) - c \sin \tfrac{1}{2}(\beta - \alpha)$$

$$= 2c \cos \frac{\pi - 2\alpha}{4} \sin \frac{\pi - 2\beta}{4}$$

$$= 4 \left(\frac{T}{g\rho} \right)^{\frac{1}{2}} \cos \frac{\pi - 2\alpha}{4} \sin \frac{\pi - 2\beta}{4}.$$

9·53. Surface tension explains the presence of a drop of liquid hanging from a body lifted out of a vessel of water. Thus if the body is a long cylinder, considering it in section, capillary curves rise to the cylinder always meeting its surface at the angle of capillarity, say at A and B. As the cylinder is lifted higher the curves approach one another until the points C, D, where their tangents are vertical, coincide; then the water breaks away and the portion above CD is left suspended as a drop.

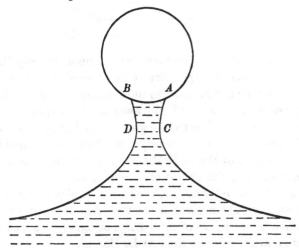

9·54. Drop of liquid on a horizontal plane. The only case capable of simple treatment is that in which the 'drop' is so large that the only curvature to be considered is that of vertical sections.

Then we have the relation of **9·4**,

$$\frac{T}{R} = p \quad \dots\dots\dots\dots\dots\dots\dots\dots(1),$$

where R is the radius of curvature of the meridian section and p is the excess of the internal over the external pressure. If we take the axis Ox along the flat top of the 'drop' and measure y downwards, the value of p at the point (x, y) on the surface is $g\rho y$, so that, putting $4T = g\rho c^2$, (1) may be written

$$Ry = \tfrac{1}{4}c^2.$$

Then if, as in **9·5**, we take ψ as in the figure, we get the same equations for the capillary curve, namely

$$y = c \sin \tfrac{1}{2}\psi,$$

and
$$x = \tfrac{1}{2}c \log \cot \tfrac{1}{4}\psi - c \cos \tfrac{1}{2}\psi + C.$$

If the origin be taken so that the axis of y touches the meridian curve, then $x = 0$ when $\psi = \tfrac{1}{2}\pi$, so that

$$x = \tfrac{1}{2}c \log \left(\cot \tfrac{1}{4}\psi \tan \tfrac{1}{8}\pi \right) - c \left(\cos \tfrac{1}{2}\psi - \frac{1}{\sqrt{2}} \right).$$

If α be the angle of contact of the liquid with the plane measured in the liquid, the height of the drop is $c \sin \tfrac{1}{2}\alpha$.

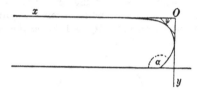

9·6. Principle of Archimedes. Effect of capillarity.

When the effect of capillarity is to depress the surface of a liquid below its normal level in the neighbourhood of a floating body, the upward force of buoyancy on the body is increased by an amount equal to the weight of liquid that would fill up the depression. In the same way if the liquid is elevated in the neighbourhood of the body the buoyancy is decreased by an amount equal to the weight of the elevation.

The volumes of these depressions and elevations in two-dimensional problems can be calculated by making use of the equations of the curve of capillarity, or instead we may make use of an artifice due to the late Dr Besant.

Consider the case of a steel needle floating upon water. Though steel is heavier than water the additional buoyancy due to capillarity will enable a needle to float if it is gently laid on the surface and not submerged.

Neglecting the end effects we may regard the problem as two-dimensional. The figure represents a section of the needle in contact with the water along the arc PAQ which subtends an angle 2θ at the centre O. E, F are the projections of P, Q on the natural level BC of the water.

The weight of unit length of the needle is supported by the

pressure on PAQ together with the surface tensions in the liquid at P and Q; and the pressure on PAQ is measured by the weight of water that would fill $EPAQF$.

Hence if α be the angle of capillarity (obtuse in this case), a the radius of the needle and 2β the angle subtended at O by

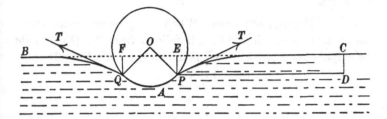

the chord of intersection by the natural level of the water, ρ the density of water and σ that of the needle, we have an equation

$$g\sigma\pi a^2 = g\rho a^2 (\theta + \sin\theta\cos\theta - 2\sin\theta\cos\beta) - 2T\sin(\theta+\alpha) \quad (1).$$

Further, considering the horizontal forces on a portion of water PDC, where DC is vertical; the horizontal tension at C balances the water thrust on CD and the horizontal component of tension at P, so that

$$T\{1 + \cos(\theta+\alpha)\} = \tfrac{1}{2}g\rho a^2 (\cos\theta - \cos\beta)^2,$$

or $\qquad 4T\cos^2\tfrac{1}{2}(\theta+\alpha) = g\rho a^2 (\cos\theta - \cos\beta)^2 \dots\dots(2).$

The elimination of T between (1) and (2) gives

$(\pi\sigma/\rho - \theta)\cos\tfrac{1}{2}(\theta+\alpha)$

$\qquad = (\cos\theta - 2\cos\beta)\sin\tfrac{1}{2}(\theta-\alpha) - \cos^2\beta\sin\tfrac{1}{2}(\theta+\alpha) \dots(3)$

as the condition of equilibrium.

9·7. Liquid films. Soap bubbles. Surface tension plays a leading part in the production of such liquid films as soap bubbles. When a loop of wire is drawn out of a soap solution a thin film of liquid is stretched across the loop. If the loop is in one plane the film is apparently plane, shewing that the action of gravity in such a film is negligible in comparison with surface tension. Both sides of the film are subject to surface tension but we shall use T to denote the sum of the tensions on the two sides of the film. On this understanding the relation between the tension of a film and the difference of the pressures on opposite sides of it is

as in **9·4.**
$$T\left(\frac{1}{r_1}+\frac{1}{r_2}\right)=p \quad \dots\dots\dots\dots\dots(1),$$

9·71. Minimal surfaces. When the pressure is the same on both sides of a liquid film, it follows from **9·7** (1) that its principal curvatures at every point must satisfy the relation

$$\frac{1}{r_1}+\frac{1}{r_2}=0 \quad \dots\dots\dots\dots\dots(1),$$

or
$$r_1=-r_2;$$

i.e. the principal curvatures are equal but in opposite senses. It can be shewn that this is the condition that the area of a surface having a given boundary may be stationary for small displacements. Taking the energy of a film as proportional to its area (**9·1**), this accords with the general proposition that a position of equilibrium is one in which the potential energy is stationary for small displacements. Surfaces which possess the property embodied in (1) are called **minimal surfaces.** The simplest is a plane.

9·72. Surface of revolution. The catenoid. Let the form of a film be a surface of revolution about the axis of x, and let the tangent to the meridian curve at any point (x, y) make an angle ψ with the axis (Fig. **9·41**). Since the pressure on opposite sides of the film is supposed to be the same, the only forces acting on any portion are the surface tensions round its boundary. Hence, by resolving parallel to Ox for the forces acting on a zone of the film between two circular sections, we get

$$2\pi y \cdot T\cos\psi=\text{const.}$$

or
$$y=c\sec\psi \quad \dots\dots\dots\dots\dots(1).$$

But $dy/ds=\sin\psi$, so that

$$\sin\psi=c\sec\psi\tan\psi\, d\psi/ds,$$

or
$$s=c\tan\psi \quad \dots\dots\dots\dots\dots(2),$$

provided we measure s and ψ so that they vanish together. Hence the meridian curve is a catenary and the surface, called a **catenoid**, is obtained by revolving a catenary about its directrix.

By reference to **9·41** we see that the general equation in this case requires that the radius of curvature R should

$= -PG$; and this is easily verified for a catenary, or conversely it can be shewn that the catenary is the only curve which possesses this property.

The meridian sections of the catenoid are convex to the axis of rotation and the circular sections are concave, making the curvatures of opposite sign.

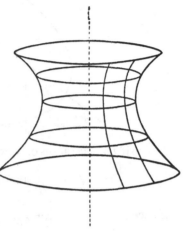

The film joining two circular rings in parallel planes on the same axis would be in the form of a catenoid.

9·73. When there is a constant difference of pressure on opposite sides of a film in the form of a surface of revolution, the condition to be satisfied is, from **9·41**,

$$\frac{1}{R} + \frac{1}{PG} = \text{const.} \quad \ldots\ldots\ldots\ldots\ldots(1),$$

where R is the radius of curvature of the meridian curve at the point P, and PG is the intercept made on the normal at P by the axis of revolution.

It can be shewn that the meridian curve is the locus of the focus of a conic which rolls on a straight line.

Let S be the focus of the conic and G its point of contact with the given line, then SG is normal to the locus of S. Let $SG = r$, and let $SY = p$ be perpendicular to the given line, and the angle $YSG = \psi$.

The (p, r) equation of a conic with the focus as origin is of the form

$$\frac{l}{p^2} - \frac{2}{r} = \mp\frac{1}{a} \text{ or zero } \quad \ldots\ldots\ldots\ldots(2),$$

according as the conic is an ellipse, hyperbola or parabola; where l is the semi-latus rectum and $2a$ the major axis.

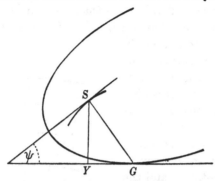

Hence if R is the radius of curvature of the locus of S, we have

$$\frac{1}{R} = -\frac{d\psi}{ds} = -\frac{d\psi}{dp}\sin\psi = \frac{d}{dp}\left(\frac{p}{r}\right) = \frac{1}{r} - \frac{p}{r^2}\frac{dr}{dp}.$$

Hence

$$\frac{1}{R} + \frac{1}{SG} = \frac{2}{r} - \frac{p}{r^2}\frac{dr}{dp}.\dots\dots\dots\dots(3).$$

But from (2)

$$\frac{l}{p^3} = \frac{1}{r^2}\frac{dr}{dp},$$

therefore

$$\frac{1}{R} + \frac{1}{SG} = \frac{2}{r} - \frac{l}{p^2},$$

and, from (2),

$$\frac{1}{R} + \frac{1}{SG} = \pm\frac{1}{a} \text{ or zero} \dots\dots\dots\dots(4),$$

according as the rolling curve is an ellipse, hyperbola or parabola.

The locus of the focus of a rolling parabola is a catenary, and a rolling ellipse and a rolling hyperbola give curves of the forms shewn below. The corresponding surfaces of revolution are called the *Catenoid*, the *Unduloid* and the *Nodoid*.

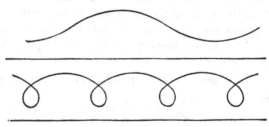

9·8. Energy of a plane film. If a plane film is drawn out from a vessel of liquid work is expended, and provided we regard the process as taking place at constant temperature this work represents the potential energy of the film.

Consider for example a rectangular film $ABCD$ bounded by straight wires AD, BC; AB being in the surface of the liquid and CD a movable wire.

The work done in pulling out the film is $T.AB.AD$, where T is the total surface tension of both sides of the film; so that if S denotes the energy per unit area

$$S = T.$$

9·9. Example. *A mass of liquid of uniform density ρ confined between two parallel planes is subject to no external forces but a constant external pressure, and the whole rotates as if solid with uniform angular velocity ω about an axis of symmetry perpendicular to the planes. Prove that, at a point on the curved surface of the liquid at a distance y from the axis,*

$$\frac{y^2+a^2}{R_1} - \frac{y^2-a^2}{R_2} = 2a \sin\alpha - c(y^2-a^2)^2,$$

where a is the radius of the circle in which the liquid meets either plane, R_1, R_2 are the principal radii of curvature, α is the (obtuse) angle of contact and $c = \rho\omega^2/4T$. [M. T.]

The equation for the pressure is

$$dp = \rho\omega^2 y\, dy,$$

giving on integration $\qquad p = C + \tfrac{1}{2}\rho\omega^2 y^2$ ....(1).

But on the boundary of the liquid

$$p - \Pi = T\left(\frac{1}{R_1} + \frac{1}{R_2}\right) \qquad ....(2),$$

where Π is the constant external pressure.

Hence $\qquad\qquad \dfrac{1}{R_1} + \dfrac{1}{R_2} = C' + 2cy^2$ ....(3).

The boundary is clearly a surface of revolution and, referring to the figure of **9·41**, we may take R_2 as the radius of curvature of the meridian curve and R_1 as the normal PG or $y \operatorname{cosec}\psi$, where ψ is the angle the normal makes with the axis.

Then measuring the arc s from A we have

$$\frac{1}{R_2} = \frac{d\psi}{ds} = \cos\psi \frac{d\psi}{dy},$$

so that (3) becomes $\quad \cos\psi \dfrac{d\psi}{dy} + \dfrac{\sin\psi}{y} = C' + 2cy^2,$

or $\qquad\qquad y\cos\psi \dfrac{d\psi}{dy} + \sin\psi = C'y + 2cy^3,$

and, on integration, $y \sin \psi = C'' + \frac{1}{2}C'y^2 + \frac{1}{4}cy^4$.

But on the plane boundary we have $y = a$ and $\psi = \pi - \alpha$, so that

$$a \sin \alpha = C'' + \frac{1}{2}C'a^2 + \frac{1}{4}ca^4,$$

and by subtraction

$$y \sin \psi = a \sin \alpha + \frac{1}{2}C'(y^2 - a^2) + \frac{1}{4}c(y^4 - a^4).$$

Putting y/R_1 for $\sin \psi$ and substituting for C' from (3) gives the required result.

EXAMPLES

1. Prove that, if a slightly conical tube of semivertical angle β is inserted in water with its axis vertical, the mean height h to which the water rises in the tube is given by

$$g\rho(ah - h^2) = 2T \cos(\alpha - \beta) \cot \beta,$$

where a is the height of the vertex of the cone above the undisturbed surface of the water, ρ the density and T the surface tension.

2. Prove that if n equal spheres of water coalesce into a single sphere the amount of surface energy liberated is proportional to $n - n^{\frac{2}{3}}$.

3. Prove that if a number of bubbles of the same surface tension coalesce into a single bubble the increase in volume bears a constant ratio to the decrease of surface.

4. Prove that, if water be introduced between two parallel plates of glass, at a distance d apart which is small compared with the linear dimensions of the film, the plates are pulled together with a force

$$AT \sin \alpha + \frac{2BT \cos \alpha}{d},$$

where A is the perimeter of the film and B its area.

5. Liquid of density ρ and surface tension T rises between two parallel vertical plates. Prove that if α is the angle of contact and h the height of the lowest point of the free surface between the plates, the height at the plates is

$$\{h^2 + 2T(1 - \sin \alpha)/g\rho\}^{\frac{1}{2}}.$$

6. Liquid of density ρ and surface tension T is in contact with a vertical plane boundary. Prove that the volume of liquid above the general level is $(T \cos \alpha)/g\rho$, where α is the angle of contact. [M. T.]

7. Obtain the result 9·6 (3) by using the equations of the capillary curve.

8. If the height of the mercury barometer is 76 cm. (specific gravity of mercury 13·6) and the tension of a soap film is 74 dynes per cm., find the percentage increase in pressure inside a bubble of radius 5 cm.

9. Two equal spherical bubbles, each of radius a, coalesce into one. Obtain an equation for the radius of the new bubble, in terms of the surface tension and the external pressure; and prove that, if the experiment were performed with bubbles of the same radius a but with greater external pressure, the new radius would be smaller. [I.]

10. A liquid is revolving about an axis under no forces except the capillary forces at the boundary, which is a surface of revolution. If the pressure at the axis is equal to that of the surrounding atmosphere, shew that the meridian curve is given by $y^3 = c^3 \sin \psi$, where y is the distance from the axis, ψ the inclination of the normal to the axis, and c is a constant.

11. A liquid film of total surface tension T is in the form of a cylinder joining two equal parallel circular discs of radius $2a$, with their centres at a distance $2a$ apart on a line perpendicular to their planes. A pinhole is made in one of the discs so that air slowly escapes. Shew that a total quantity

$$\pi \rho_0 a \left[8a^2 \{1 + T(2a\Pi)\} - c^2 \{1 + 2 \sinh(a/c)\} \right]$$

will escape, where ρ_0 and Π are the atmospheric density and pressure, and c is given by $\cosh(a/c) = 2a/c$. [M. T.]

12. A tube is drawn out into a point, in such a way that the internal cross-section at a distance y from the point is a circle of radius y^2/a. The tube is held in a vertical position with the point downwards, and a volume V of liquid of density ρ is poured in. Assuming that both bounding surfaces of the liquid are spheres of radii small compared with their distances from the point, and that the air pressure on both surfaces is the same, shew that the distances y_1, y_2 of the two bounding surfaces from the point are given by the equations

$$V = \tfrac{1}{3}\pi \frac{y_1^5 - y_2^5}{a^2}, \quad 2aT \cos\alpha \frac{y_1 + y_2}{y_1^2 y_2^2} = \rho g,$$

where $\pi - \alpha$ is the angle of contact. [M. T.]

ANSWERS

8. 0·0029.

9. $2T(r^2 - 2a^2) = \Pi(2a^3 - r^3)$, where Π is the external pressure.

Printed in the United States
By Bookmasters